Forestry Pesticide Aerial Spraying

Environmental Science and Technology Library

VOLUME 12

The titles published in this series are listed at the end of this volume.

Forestry Pesticide Aerial Spraying

Spray Droplet Generation, Dispersion, and Deposition

by

J.J.C. Picot

Department of Chemical Engineering,
University of New Brunswick,
Fredericton, NB, Canada

and

D.D. Kristmanson

formerly of
Department of Chemical Engineering,
University of New Brunswick,
Fredericton, NB, Canada

SPRINGER-SCIENCE+BUSINESS MEDIA, B.V.

Library of Congress Cataloging-in-Publication Data

Picot, J. J. C.
Forestry pesticide aerial spraying : spray droplet generation, dispersion, and deposition / by J.J.C. Picot and D.D. Kristmanson.
p. cm. -- (Environmental science and technology library ; v. 12)
Includes index.
ISBN 978-0-7923-4371-4 ISBN 978-94-011-5634-9 (eBook)
DOI 10.1007/978-94-011-5634-9
1. Aerial spraying and dusting in forestry--New Brunswick.
2. Forest insects--Control--New Brunswick. 3. Aerial spraying and dusting in forestry. 4. Forest insects--Control. I. Kristmanson, D. D. II. Title. III. Series.
SB764.C36P53 1997
634.9'694--dc21 96-37439

ISBN 978-0-7923-4371-4

Printed on acid-free paper

TABLE OF CONTENTS

PREFACE

Introduced to the technical aspects of forestry aerial spraying in the mid-1970's, we were immediately impressed by the complexity of the process of delivering pesticide to foliage. At that time, there was a vigorous public debate in New Brunswick about the ecological and public health impacts of the annual spray program for the control of defoliation of spruce and fir trees by the spruce budworm. The forest industry is important to the province and changes to the established procedures of budworm control could have major economic implications. A rational debate required reliable information about the mechanics of the spraying process. There was a need to supply missing information as to required pesticide application rates, atomizer performance, off-target drift and deposit, and the effects of weather and aircraft operating factors. We were invited to initiate a research program in this domain by New Brunswick forest management officials, and what follows in this book is a logical and quantitative description of the overall process based on our own research and that of others over the intervening years.

After a short introduction to aerial spraying, we begin (Chapter 2) by describing forest stands in terms of their interaction with suspended atmospheric particulate material carried along by the wind and susceptible to deposition on foliage. We introduce foliage simulators and their use in measuring the deposit of sprayed pesticide on foliage, the "biological interface" between pest and pesticide. Chapter 3 treats the relationship between toxic requirement for pest control and droplet size, and the effects of physico-chemical properties of the formulation on droplet size. In Chapter 4, in addition to a cursory discussion of spray aircraft types and wake behaviour, we present a detailed account of spray atomizer droplet characterization methods and equipment, with an innovative use of the Malvern droplet spectrometer for flux-based droplet spectrum measurements. Tables of droplet spectrum data are given in Appendix II.

Chapter 5 summarizes spray field trial methods and management, with a discussion of the importance of the mass balance criterion in assessing the overall accuracy of results.

All of the mechanisms of the spray generation and deposition process are brought together within the framework of the PKBW computer model in Chapter 6. This is a droplet trajectory model permitting the calculation of deposited droplet size and density on the biological interface, as well as size and concentration in the drift cloud. Long-distance drift calculations are presented, as are model validation results.

The last three chapters cover recent USDA Forest Service research on spraying in complex terrain, a presentation of some recent spraying innovations, and a review of some government regulations relating to pesticide application.

We claim here to have a unique quantitative treatment of the many mechanisms involved in the pesticide application process. With the methods described here, it is possible to get a detailed and accurate prediction, via computer simulation, of droplet deposition and downwind drift of pesticide droplets. This offers an unparalleled tool in the continued development of pesticide application technology.

We strongly encourage scientists, regulators, and forest managers to use the technology presented here in the interest of environmentally conscious forest management.

J.J.C. Picot, Fredericton, N.B.
D.D. Kristmanson, Surrey, B.C.

ACKNOWLEDGEMENTS

We recognize here the central role played by Mr. H.J. (Bud) Irving, then Managing Director of Forest Protection Ltd. of Fredericton, N.B., in initiating our interest in spray technology research and continuing as a sponsor and mentor of our research activities over the years. Acting on the advice of Dr. Frank Webb, former Director of the Maritimes Forest Research Center, he was successful in getting the collaboration of scientists from a wide range of disciplines in addressing problems in pesticide application technology soon after assuming his responsibilities in 1975.

Our early learning experiences in this domain were greatly enhanced by interaction with colleagues already active in various aspects of spray technology research. Principal among these were R.S Crabbe of the National Research Council (NRC), Ottawa (dispersion of atmospheric pollutants and field trials); John Reid, Environment Canada, Ottawa (modelling the transport of atmospheric particulates); W.I. Varty, Forestry Canada, Fredericton (first chairman of interdisciplinary committee on monitoring of spray drift in N.B.); R.H. Wickens, NRC, Ottawa (wind tunnel operation and design, aircraft vortex behaviour); and Charles Wiesner, N.B. Research and Productivity Council (RPC), Fredericton, N.B. (first coordinator of the New Brunswick Spray Efficacy Research Group (now known as SERG)). Later, we were given valuable assistance by J.W. Barry, USDA Forest Service, Davis, CA, and J.E. Hirsh, Princeton Aeronautical Research Associates, Princeton, N.J. in the area of aircraft spray transport and wake modelling. We greatly benefited from discussions with Professors Norman Akesson and Wesley Yates of the University of California at Davis, and earlier with R. Courshee, S. Parkin, and V. Joyce of the Cranfield Institute of Technology, Cranfield, England.

Our research team at the University of New Brunswick has included many talented and dedicated individuals. Mr. Alfred Knappe, Chief Technician in our Chemical Engineering Department, was responsible for constructing the SERG/ONAIR prototype atomizers, and our wind tunnel with his assistant Brian Hawkins. Excellent work was done by the following graduate students over the years: Bruce Smedley, Terry Tomney, Malcolm Thompson, Xavier Bontemps, Nusin Basak-Brown, Michael Pang, David Wallace, and Robert Goguen. Stability and continuity in our wind tunnel and field trial work were assured by the efforts of Research Engineer Mark van Vliet assisted by Gary Henderson, Research Technician.

In recent years, our work has greatly benefited from collaboration with R.E. Mickle, Environment Canada, Downsview, Ont., D.C. Davies, Forest Protection Ltd., Fredericton, N.B., E.G. Kettela, Canadian Forest Service, Fredericton, N.B., and C.M. Riley, RPC, Fredericton, N.B.

We are acutely aware that over twenty years we have learned from many others not acknowledged here about practical and theoretical matters and have enjoyed the co-operation, advice, and support of many participants in SERG activities. Those we also thank.

We are grateful for the research funding we have received from the National Research Council; Forest Protection Ltd.; the Natural Sciences and Engineering Research Council, Canada; the Atmospheric Environment Service Science Subvention Program; the Canadian Forestry Service; and the University of New Brunswick. Also, we thank Dr. J.A. Zabkiewicz of the New Zealand Forest Research Institute for arranging a useful collaboration with JJCP which, among other things, resulted in the launching of the manuscript for this text.

We thank Ms. Laurie Nason for her work in the typing and preparation of the text for publication.

Lastly, we endorse the employment policies of the University of New Brunswick which have allowed us to devote a significant portion of our time to this regionally important forest management problem.

CHAPTER 1

INTRODUCTION

1.1. Historical Perspective

Pesticide application by aircraft has been used in forest management for many years. Table 1 shows data for extent of defoliation and area treated for spruce budworms in North America in 1983 and 1984 [Kucera and Taylor(1985)]. According to preliminary data compiled for 1994 for the United States, 354,000 ha of forest were defoliated by budworms in 1994 [Kucera(1994)], with no treatment applied. The North American gypsy moth accounted for 348,000 ha of defoliation, 183,000 of it in Virginia, where 46,000 ha were treated by aircraft. In Canada, 168,277 ha were treated by aircraft in 1993[Can. Coun. of For. Min. (1995)], for all insects. In New Brunswick, aerial spraying to control defoliation by the eastern spruce budworm (Choristoneura fumiferana(Clem.)) started in 1952, rose to a maximum area treated of 3.9 million ha in 1976, and declined to 0.1 ha in 1993, with no treatment in 1994 [Kettela(1995), Davies(1995)].

Herbicide application to control unwanted plant species was carried out on 42,000 ha of forest land in the United States in 1994, and on 139,000 ha in Canada in 1993, according to the above reports. Ground application accounted for most of the U. S. program, while aerial application was most often used in Canada.

In New Zealand, aerial spraying for pine needle blight was carried out over a total of 378,000 ha in the period 1990 to 1995, while herbicide spraying in radiata pine plantations was done over a total of 245,000 ha in the period 1991 to 1994 [Richardson(1995)].

In the Czech Republic, about 15,000 hectares of forest are treated with pesticides annually [Linkova (1991)].

Public concern about the need to maintain a healthy global environment has led governments to develop strategies for integrated resource management. In Canada, Natural Resources Canada has launched the Integrated Forest Pest Management Initiative, which encourages researchers to "develop ecological approaches to maintaining pest-caused damage at economically acceptable levels through knowledgeable, integrated use of a variety of pest management options" [Caldwell(1995)]. Pesticide spraying is one of these options.

Our interest in spraying problems was initiated by a request to examine the mechanics of the spraying process during a severe episode of defoliation of fir and spruce trees by spruce budworms in New Brunswick in the mid-1970's. Other scientists and engineers in Eastern Canada were drawn into this research effort and an umbrella organization, the Spray Efficacy Research Group (SERG), soon emerged to co-ordinate the efforts of workers drawn from several provincial and federal government agencies and from universities (see Appendix I). Forest Protection Limited, an organization which was funded by the NB government and the forest industry and charged with the task of running spruce budworm spraying operations, was largely instrumental in starting SERG and maintaining interest and financial support over many years. Headquartered in Fredericton, NB, SERG is dedicated to improving the efficiency of pesticide application so that the amount of pesticide needed to manage forest pests is reduced to an absolute minimum [Irving(1994)].

Much of the early SERG field trial work was done at a test site near Dunphy, N.B. situated at 46.580N, 65.830W. This work will be extensively referred to in this book as "the Dunphy trials".

TABLE 1.1. Defoliation caused by spruce budworms (Choristoneura spp.) In North America in 1983 and 1984 (Kucera and Taylor (1985))

Province, State, or Forest Service Region	1983	1984	Trend* in 1984
	1,000 hectares		
Newfoundland	67	15	52 -
Nova Scotia	358	85	273 -
Prince Edward Island	26	15	11 -
New Brunswick	2,329	730	1,599 -
Quebec	12,266	11,200	1,066 -
Ontario	9,033	8,748	285 -
Alberta, Manitoba, and Saskatchewan	76	170	94 +
Maine	2,430	2,228	202 -
New Hampshire	2	< 1	1 -
Vermont	72	0	72 -
Michigan	60	78	18 +
Minnesota	51	147	96 +
Wisconsin	9	9	0
British Columbia	239	74	165 -
Northern Region	1,059	900	159 -
Rocky Mountain Region	1,100	888	212 -
Southwestern Region	140	280	149 +
Intermountain Region	1,127	950	177 -
Pacific Southwest Region	42	53	11 +
Pacific Northwest Region	1,047	1,385	191 +
Alaska Region	1	0	1
Total	31,534	27,956	3,716 -

* - = decrease in defoliation.
\+ = increase in defoliation.

The United States Department of Agriculture Forest Service is also developing a Tactical Plan for seed, cone, and regeneration insect control in the context of integrated pest management [Barry(1993)].

1.2. Scope and Orientation of Text

This is a reference text explaining the application of pesticides to vegetation using aircraft. Each stage of the process is examined in detail, i.e., particle generation by atomizers, the effect of aircraft wake on the cloud of particles, atmospheric dispersion of the cloud, evaporation of material from individual particles, the interaction of the particles with vegetation, and the eventual deposition of the particles. These factors are covered in the first five chapters, and the sixth describes how they are combined to produce a computer simulation of the overall process.

The objective is to provide forest managers and regulatory officials and their engineers and technicians with the necessary data and calculation methods to carry out systematic predictions or evaluations of proposed or actual forestry spray operations. Toxicity is not treated per se, but the reader will be able to determine dosage and deposition rates in and near treated areas. With toxicity data for a specific formulation obtained elsewhere, toxic effect can then be estimated. The methods mentioned are set up in a generic fashion, to permit effective use of the text.

References

Barry, J.W. (1993) Fifth report: National Steering Committee for Management of Seed, Cone, and Regeneration Insects, FPM 93-13, USDA Forest Service, Forest Pest Management, Davis, Ca.

Caldwell, E. (1995) publicity sheet on Integrated Forest Pest Management Initiative, Canada's Green Plan, Natural Resources Canada, P.O. Box 490, Sault Ste. Marie, Ontario.

Canadian Council of Forest Ministers (1995) The Compendium of Canadian Forestry Statistics (1994), National Forestry Database, Ottawa.

Davies, D. (1995) General Manager, Forest Protection Ltd., Fredericton, New Brunswick, personal communication.

Irving, H.J. (1994) Coordinator, Spray Efficacy Research Group (SERG), 79 Silverwood Terrace, Fredericton, N.B., personal communication.

Kettela, E.G. (1995) Insect control in New Brunswick, Chapter 64 in Forest Insect Pests in Canada, J.A. Armstrong and W.G.H. Ives eds., Natural Resources Canada, Canadian Forest Service, Science and Sustainable Development Directorate, Ottawa.

Kucera, D.R., and R.G. Taylor (1985) Spruce Budworms Situation in North America 1984, U.S.D.A. Forest Service Misc. Pub. #1448.

Kucera, D.R. (1994) Insect and disease conditions in the United States, prepared for the 22nd annual Forest Pest Control Forum, Nov. 15-18, Ottawa, Canada; Northeastern Area State and Private Forestry, U.S.D.A. Forest Service, Radnor, Pennsylvania.

Linkova, E. (1991): ed., Agriculture in the Czech Republic, #742, Press and Information Centre, Ministry of Agriculture of the Czech Republic, Park CENTRUM Ceske Budejovice, CR.

Richardson, B. (1995), New Zealand Forest Research Institute, Plant Protection Chemistry division, Rotorua, N.Z., personal communication.

CHAPTER 2

CHARACTERIZING FOREST STANDS

The characteristics of a forest plot which are important in the context of the deposition of particles dispersed in the atmosphere will be described. These are 1) foliage density and its distribution, 2) micrometeorology, 3) the capture of airborne particles by foliage, and 4) the biological interface to be targeted. The use of foliage simulators in monitoring spray application is also addressed in this chapter.

2.1 Foliage Density and Vertical Distribution

Most of the insecticides used to protect against defoliating insects have a direct contact mode of action, either by external pickup or by ingestion by larvae, or both. The conventional approach is to apply the insecticide to the forest in the period when insects are active enough to intercept lethal doses of insecticide but before an overly damaging amount of foliage has been consumed. The target for insecticide droplets is therefore the foliage. In the case of herbicide application, a phytotoxic material is often used whose mode of action is to do sufficient damage to the foliage to kill the plant. Again, the foliage is the target. The total quantity of foliage in a plot to be sprayed is, therefore, of prime importance in designing a pesticide application.

In a paper describing the measurement of foliage and ground deposits for the Dunphy spray trials, Picot et al. (1993) give foliage density and distribution data from measurements done at the site in 1982 (Table 2.1 and Figure 2.1). This forest was a naturally-regenerated stand in a previously cut-over area. The data represent total foliage density (as m^2 per m^2 of ground) and its vertical distribution within a test block oriented downwind of an aerial spray application line or flight path.

The procedure followed to obtain the data was as follows: At each downwind location, shown in Table 2.1, 0.010ha plots were surveyed by measuring all tree trunk diameters at breast (technician's) height (DBH). The species present were black spruce (Picea mariana (Mill.)B.S.P.), jackpine (Pinus banksiana(Lamb)), and balsam fir (Abies balsamea(L.)). On-site representative specimens of these were cut down in order to get data on the relationship between breast-height diameter and height as well as on the weight of foliage as a function of height. Foliage was stripped from the whole length of the tree, at one meter intervals, and weighed, and these weights eventually totalled. Five-kilogram representative samples of each species were placcd in sealed plastic bags at the site and taken to the laboratory where samples of needles were removed and weighed and measurements taken of the needle areas (maximum projected area of the needles on a plane parallel to the long axis of the needle). Since the original

TABLE 2.1 Stand characteristics: Tree counts and average heights by species and foliage area for 10 plots along the sampling line [Picot et al (1993)]

Distance Along Sampling LIne (m)	Balsam Fir Height (m)	Balsam Fir Stems per Ha	Spruce Height (m)	Spruce Stems per Ha	Jack Pine Height (m)	Jack Pine Stems per Ha	Foliage Area* (total) m^2/m^2 ground
50	4.9	173	4.2	688	11.1	855	2.64
100	5.9	81	4.8	1171	9.4	702	2.99
150	5.9	91	4.9	1293	10.8	417	3.11
200	6.3	20	5.3	671	9.2	906	2.93
250	4.3	40	4.8	1192	9.1	753	2.81
300	6.0	102	5.6	804	14.3	336	2.82
500	7.3	30	5.3	2382	12.6	204	3.41
600	6.4	71	6.1	2209	12.2	61	4.32
800	5.7	71	5.8	1456	12.7	153	4.01
Average	5.9	80	5.1	1304	11.0	492	3.2

* Foliage area is defined as the maximum projected area of the needles normal to the long axis of the needle

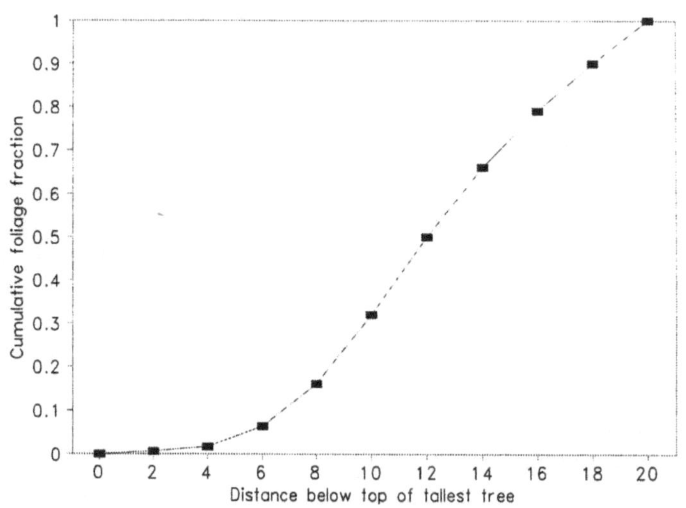

Figure 2.1. Average accumulated foliage distribution, Dunphy site

weighed foliage samples consisted of needles and twig stems, the mass ratio of foliage to twig stems was found by weighing. These data were converted to foliage area for each tree sampled. Also, foliage area as a function of height for each tree species was found [van Vliet(1987)] and is shown graphically for jackpine and black spruce on Figure 2.2, along with values from the Ker(1984) equations which are based on samples taken from 200 sites distributed throughout New Brunswick and Nova Scotia. In addition, the cumulative foliage was found as a function of distance from the top of the tree from the data on each tree of each species [Picot (1986)], and averaged to give the vertical distribution patterns such as shown on Figure 2.2, for each species. The total foliage per square meter of ground and the vertical foliage distribution in the plot were then calculated as follows:

1. Foliage area for each tree was found from the graphs of Figure 2.2 for the appropriate species. Values for all trees were added to give the total foliage area and the foliage area index (foliage area per square meter of ground) for each .010ha block. An average of these gives the representative value for foliage area index for the test site.
2. For each tree in each of these .010ha blocks, the quantity of foliage in each 2-meter height increment was calculated using the Figure 2.2 data for area vs height of tree and for vertical distribution of that area, according to the species.
3. All areas for each 2-meter height increment in the test site were added to give the average vertical distribution of the area for the test site.
4. The data from step 3. were converted to cumulative foliage fraction versus percent of distance from the top of the tallest trees in the stand, as shown on Figure 2.1. This plot along with the average foliage area index of Table 2.1 can then be used to compute the area of foliage between any two vertical positions in the canopy, per square meter of ground area, for the test site.

When data of the type mentioned above are not specifically available, it is possible to estimate foliage area index from surveys of tree type and DBH and correlations for foliage biomass vs DBH such as those of Ker (1984). The relation between biomass and foliage area can be found by measuring all needles of a weighed sample. (The values given by van Vliet (1987) for jackpine, black spruce, and balsam fir are 18.6, 15.5, and 15.3 sq. cm per gram, respectively). Information about vertical distribution pattern is found in Belot, et al (1976), and on-site inspection and limited sampling can provide this with modest effort. It is worth noting that significant success has been achieved using airborne imaging synthetic aperture radar in determining biomass components in test forests [Kasischke (1995), Sun and Ranson (1995)]. Also, Grim et al. (1992) and Steinke et al (1996) refer to a photometric instrument for making non-destructive measurements of

Jackpine:

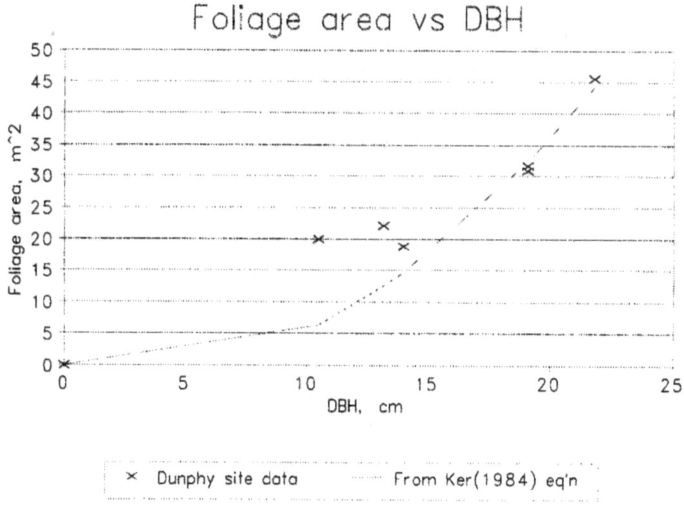

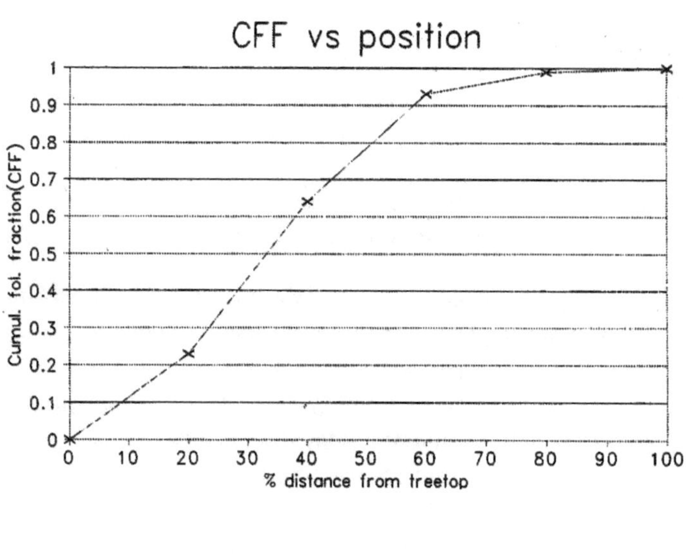

Figure 2.2 Foliage area vs tree height

Black spruce:

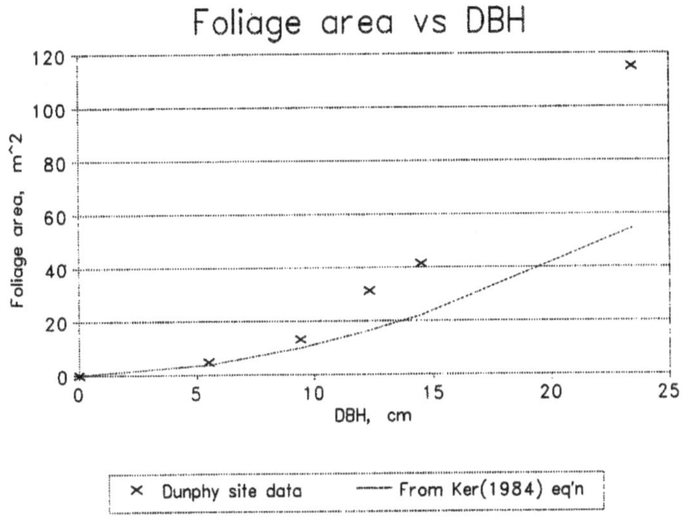

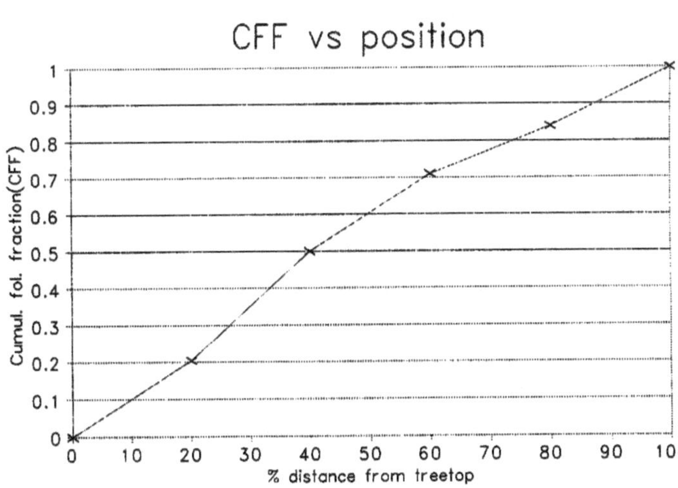

Figure 2.2 Continued

leaf area index. A recent initiative called the International Canopy Network (ICAN) headquartered at The Evergreen State College, Olympia, WA 98505, publishes a newsletter which contains a citation list of publications relating to canopy properties.

The development of broadleaf species foliage in a conifer plantation is monitored before application of herbicide. A test site near Sevogle, N.B. (47° 05'N, 66°06'W) was clearcut in 1988 and replanted in1990 and was due to be herbicided in 1992. The test project was carried out in 1991 when the average broadleaf area index was $0.48m^2$ per m^2 of ground with a maximum height of 2.5m [Riley, C.M. (1992), Chapman, T. (1993)].

The utility of the above data will be explained in detail in Chapter 6 where modelling of pesticide deposition will be done by calculating individual droplet trajectories from the spray aircraft into the canopy and onto the foliage.

2.2 Forest Micrometeorology

Wind velocity profiles over forested ground are described in detail in texts by Geiger(1966), Rosenberg(1974), and Monteith(1973). Correlations which depend on atmospheric stability are provided. The stability of the atmosphere depends on the relation between the actual vertical temperature distribution and the adiabatic temperature distribution. The adiabatic distribution is calculated by assuming that the air temperature at any height T_h is related to that at the surface T_s by an adiabatic reversible expansion to the pressure existing at that height, according to:

$$T_h = T_s \left(\frac{P_h}{P_s} \right)^{\frac{R}{C_p}} \tag{2.2.1}$$

This relation will produce a gradual reduction of temperature with increasing height equal to 0.01° C per meter, called the dry adiabatic lapse rate. (P_h and P_s are pressure at height and at surface and R is the universal gas constant while C_p is the air heat capacity at constant pressure). Such an atmospheric temperature distribution ensures that neutral buoyancy exists for a parcel of air which is displaced vertically up or down by some disturbance. That is, the parcel will remain where it has been moved once the disturbance has disappeared. When the temperature gradient is greater (less negative and gradually more positive), the atmosphere is stable in the sense that the air parcel would move back towards its original position. For a more negative gradient, the atmosphere is unstable and tends to mix vertically spontaneously, i.e. the vertical displacement of the air parcel tends to increase. For a neutral atmosphere situation, for wind over a large, flat, uniform surface (forested or otherwise) the mean velocity profile can be described by the logarithmic law:

$$u = \frac{u^*}{k} \ln\left(\frac{z - z_d}{z_o}\right) \tag{2.2.2}$$

Here, u is the time-averaged horizontal wind velocity component, u* is the characteristic wind velocity known as the friction or shear velocity, z the height above ground, z_d the displacement height, z_o the roughness height, and k the von Karman constant with a value of 0.4. The three constants are found experimentally from time-averaged values of velocity measured simultaneously from at least three different vertical positions for the location of interest, substituting the three sets of height and velocity values into equation 2.2.2 consecutively to obtain three independent equations for the three constants. For the Dunphy, N.B. test site mentioned in the above section, Crabbe et al. (1983) give a z_d of 11.3m and a z_o of 1.44m. Measurements were taken from an instrumented meteorological tower set up in the forest area. Belot et al.(1976) give values from six other sources which show an average ratio of 0.76 for z_d over mean tree height, and 0.064 for z_o over mean tree height. Bergen(1979) gives values of 0.88 and 0.015 respectively for these ratios for a mixed Douglas fir and western larch stand. The Dunphy ratios are 1.18 and 0.15.

For stable and unstable atmospheres, the above velocity profile equation is modified as follows:

$$u = \frac{u^*}{k} \ln\left(\frac{z - z_d}{z_o}\right) \phi \tag{2.2.3}$$

where $\phi = (1 + Ri)^{.333}$ for stable cases, and
$\phi = (1 - Ri)^{-.333}$ for unstable cases.

Here, Ri is the dimensionless gradient Richardson number:

$$Ri = g\left(\frac{d\Theta}{dz}\right)\left(T_h\left(\frac{du}{dz}\right)^2\right)^{-1} \tag{2.2.4}$$

where g is the gravitational constant, du/dz is vertical velocity gradient, and the potential temperature gradient $d\Theta/dz$ is the sum of the actual measured temperature gradient dT/dz and the adiabatic lapse rate -.01°C per meter [Seinfeld (1975)]. That author also discusses the effect of high humidity on lapse rate. The Ri number is an index of the combined influences of temperature gradient and wind speed in generating turbulence in the atmosphere. Time-averaged readings of velocity and temperature at different heights are needed to calculate the gradients needed in the definition. The Richardson number is negative for unstable, positive for stable, and zero for neutral atmospheres.

Figure 2.3 a), b), c), d) show measured temperature profiles for Dunphy spray trials and the changes occurring from sunrise through mid-day and on into the evening, as solar radiation increases and decreases. In 2.3 a) an early-morning stable layer is shown below the neutral layer aloft, while 2.3 b) shows developing instability near the surface caused by solar heating. Well-developed instability is shown for afternoon conditions in 2.3 c), and a redeveloping stable layer caused by evening cooling is shown in 2.3 d). As the temperature profile moves past the neutral into the unstable condition, the buoyancy of lower-lying air forces it to move upwards, creating turbulence which promotes vertical mixing. The zone involved in the mixing increases in thickness as the day progresses until late afternoon and evening when the earth's surface starts cooling. This surface cooling causes a cooled air layer to develop at ground level, and the thickness of this layer increases through the night until dawn the following morning when the whole process repeats itself, assuming cloudless skies. The increased degree of mixing caused by increasing solar radiation is reflected in an increased degree of frictional drag (increased turbulent viscosity) between one air layer and the next, when a wind exists aloft. For a constant wind speed at the top of the mixing layer, the result is low wind speed near the ground in the morning, followed by increasing winds, until afternoon and evening cooling cause a reversal in the process. Note that wind velocity itself can create (mechanical) turbulence, a normal manifestation of flow as velocity increases. The Monin-Obukhov length is used to express a characteristic height at which turbulence effects from both mechanical and thermal sources are equivalent:

$$L = -\frac{u^{*}\,T}{k\,g\left(\dfrac{dT}{du}\right)} \tag{2.2.5}$$

Here dT/du is the gradient of temperature with respect to wind velocity obtained from measurents at various heights. The Monin-Obukhov length is negative for unstable, positive for stable, and has the value infinity for neutral atmospheres.

Raynor(1971) has measured wind transition effects for a 10.5 m high pine forest contiguous to an open field. For winds blowing into the forest, profiles gradually change to reflect the upward displacement of the lower boundary (i.e. higher values of z_d and z_o) in the forest, for a distance into the forest of about four

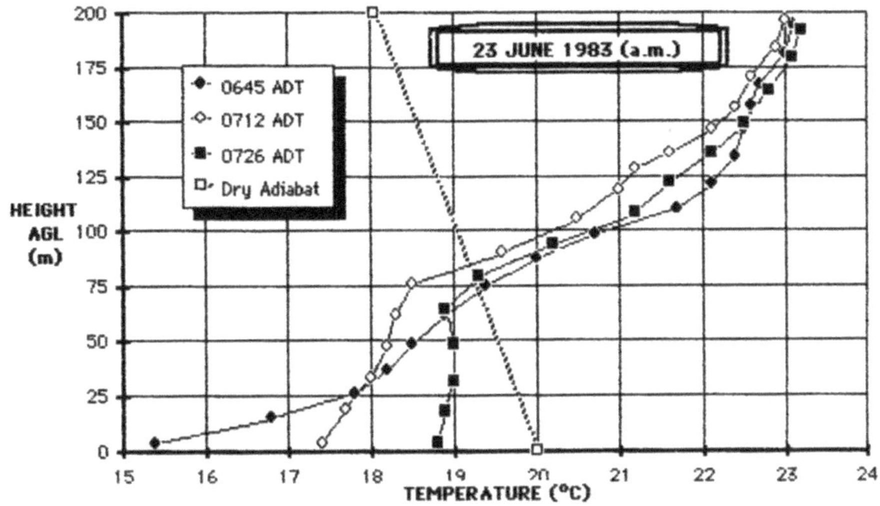

a) Early morning strongly stable atmosphere, Dunphy, N.B. test site showing beginning of surface heating after sunrise.

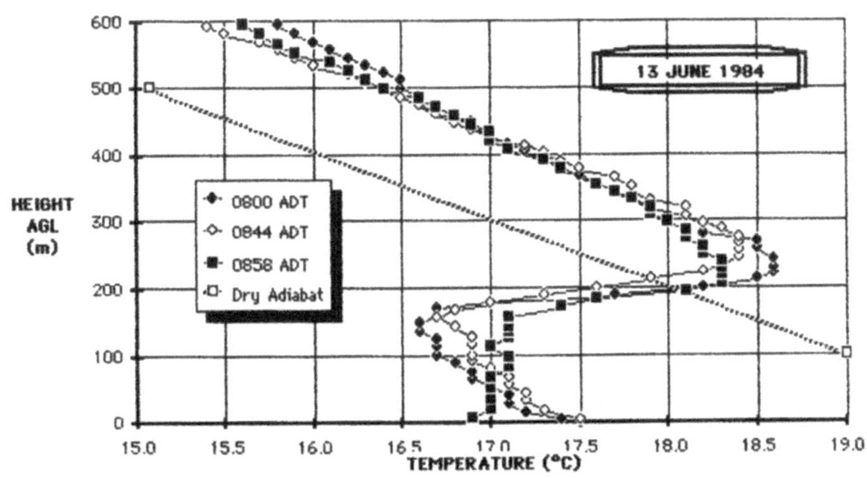

b) Continuation of early morning surface heating under very stable layer.

Figure 2.3 Temperature as function of height above ground level, Dunphy test site, copied from Steeves and Dickison (1987) by permission

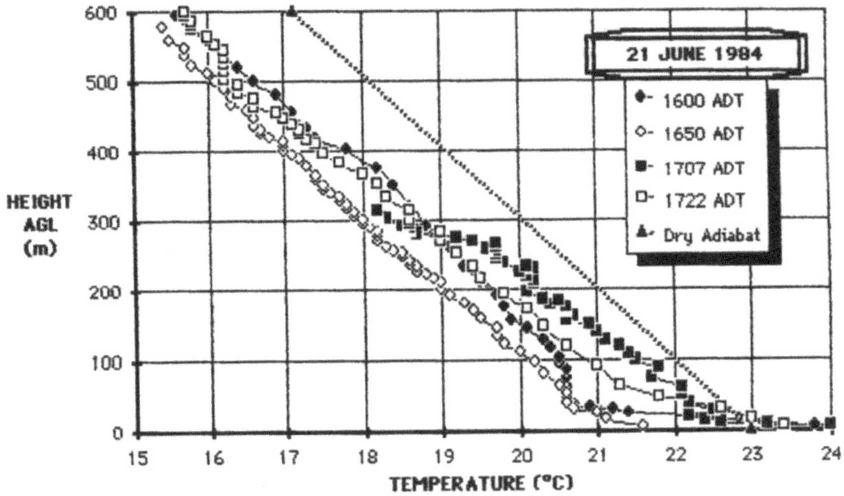

c) Unstable conditions in the afternoon, with stable layer dissipated by thermal mixing

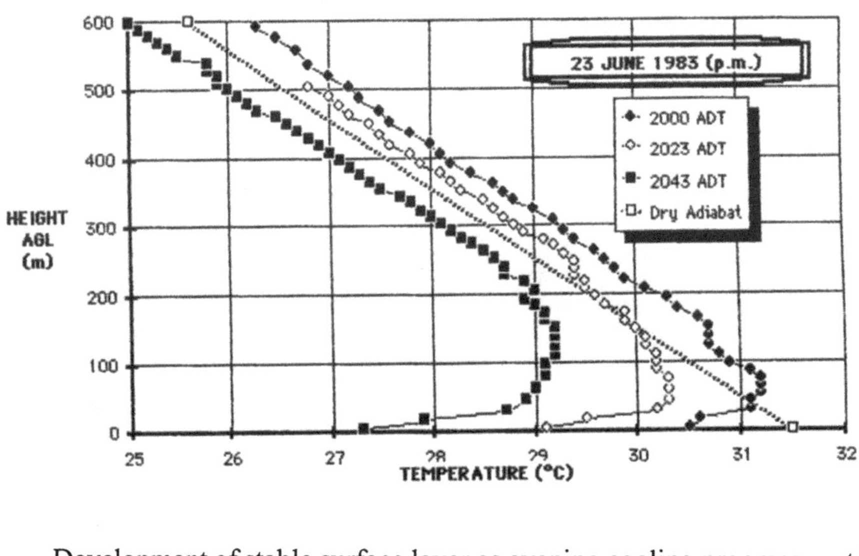

d) Development of stable surface layer as evening cooling progresses at low level.

Figure 2.3 Continued

times tree height. He also shows that the development of sub-canopy profiles is also essentially complete after a distance of six times tree height into the forest.

Wallace et al. (1995) recommend the following generally accepted expression for canopy flows be used with a value of 2.5 for the constant α for all stability classes for the fully developed profile:

$$u(z) = u(z_T) \exp\left[-\alpha\left(1 - \frac{z}{z_T}\right)\right] \qquad (2.2.6)$$

Here, z_t is mean tree height.

In Chapter 6 the above equations will be used in modelling droplet transport and deposition for aerial spraying operations along with correlations for atmospheric turbulence which will be explained there.

2.3. Foliage as Filter for Amospheric Particulates

Pesticide aerial application in forestry involves liquid droplets or solid granules which most often are intended to be applied directly to the foliage. Liquid droplets are produced by atomizers which generate a spectrum of droplet sizes with droplet frequency (or counts) being inversely proportional to droplet size (see section 6.2). That is, the smaller the droplet, the more there are of them in the spray. Atomizers producing a smaller average size have a narrower spectrum than those producing a larger average size. Insecticide spraying for spruce budworm is done with sprays with average droplet size well below 100 micrometers. Droplets in this range have low settling velocities (see Figure 2.4, e.g. a 50 micrometer droplet has a settling velocity in air of 7.5cm/s in air at 20° C). Crabbe et al (1980 and 1983) have shown that droplets smaller than 20 micrometers will persist for a long time in the drift cloud carried downwind from a spray application, but that, nevertheless, an exponential decay of droplet concentration with distance occurs. Wind speeds in a forest canopy vary from near zero to several meters per second. Of the four mechanisms of aerosol collection, namely, direct interception by gravitational settling, inertial impaction, Brownian diffusion, and electrostatic attraction, the first two are of significance in this case [Strauss (1975)]. Inertial impaction is the phenomenon which allows small diameter foliage elements (e.g. needles) to collect droplets suspended in and moving past in an air stream when these same droplets will be carried past a larger foliage element by the air flow which is diverted around the obstacle. The tendency to flow around the obstacle is promoted by smaller droplet size and larger diameter obstacles, whereas the tendency to strike the obstacle is promoted by larger droplets and smaller obstacles. Other things being equal, higher air velocities promote greater tendency to deposit on the obstacle.

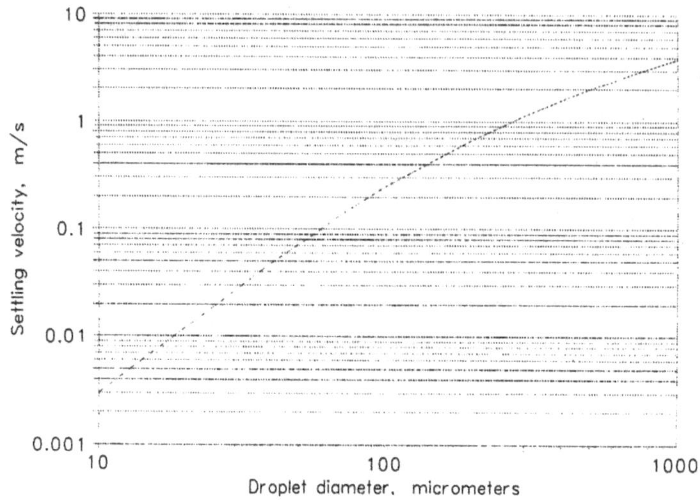

Figure 2.4 Droplet settling velocity in air at 20°C [Strauss (1975)]

The impaction of droplets on obstacles of simple geometry (cylinders, spheres, ribbons) was analyzed by Langmuir and Blodgett (1946) and Davies and Peetz (1956). May and Clifford (1967) gave experimental results showing slightly different values, but their values for ribbons are the ones chosen here. The fraction of the droplets in the stream approaching an obstacle which deposits correlates with the Stokes number, defined as:

$$S_t = \frac{u\ V}{g\ D} \tag{2.3.1}$$

where V = terminal settling velocity of the droplet,
u = air approach velocity,
D = width of ribbon or cylinder,
and g = gravitational constant.

The collection efficiency results of May and Clifford (1967) for ribbon-like obstacles are given below using a curve-fit for their experimental data:

$$\eta = S_t^3/(S_t^3 + 1.1657\ S_t^2 - 0.5695\ S_t + 0.2362) \tag{2.3.2}$$

Collection efficiency η is the number of droplets collecting on the obstacle divided by the total number of droplets approaching the obstacle on an intercepting path.

Assuming that a balsam fir needle can be represented by a ribbon about 0.2 cm wide, and that droplets are carried along in an air stream moving at 0.5, 1.0, and 2.0 m/s at 20°C at normal atmospheric pressure, Figure 2.5 shows efficiency values for droplets of different sizes calculated from equation 2.3.2.

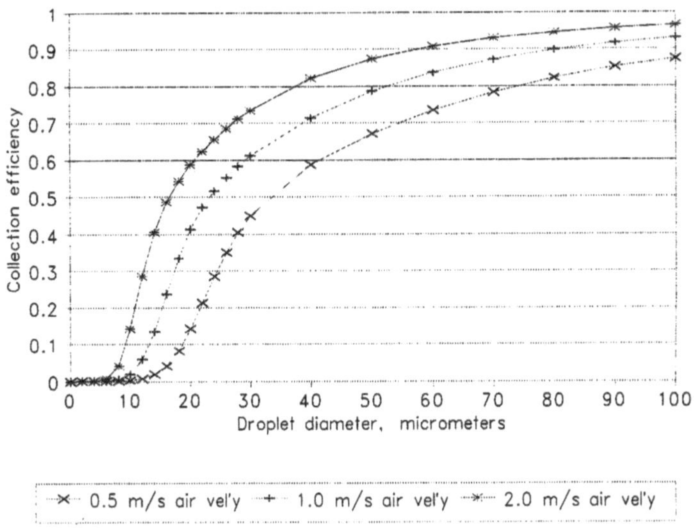

Figure 2.5. Collection efficiency for water droplets on 2mm wide ribbon, airflow perpendicular to the flat side

Suspended droplets 15 micrometers and larger will collect with significant efficiency on individual balsam fir needles in the direct path of a flow of air at velocities near 1.0m/s.

The following factors are also important:

1) Some of the approaching flow will be diverted around the shoot because of its porous nature. The velocity of approach of particles to needles will be less than the velocity upstream of the shoot. In other words, the Stokes number of the actual encounter cannot be calculated using the velocity upstream of the shoot or tree. In general, the collection efficiency will be lower.

2) The matrix of shoots, twigs, and branches, and trunk creates a complex interaction between the air and droplets moving through it and the individual collecting elements. Some of the droplets suspended in the air flow will pass through the tree without encountering a needle while others will undergo one or more encounters before depositing on a needle. Upstream shoots will collect more droplets than downstream ones. Encounters will also be

complicated by the random orientation of the needles.

These considerations are consistent with the reports of Wedding et al.(1978) on the distribution of droplets on Douglas fir, and with the observations of Armstrong and Yule (1978) on the filtering effect of fir and spruce foliage on pesticide distribution on the downwind side of trees. Kristmanson et al. (1984) studied item 1) above with single fir, spruce, and jackpine shoots suspended at various orientations to air flow direction in an air flow test section having a measured air velocity (Figure 2.6). Droplets were generated with a Berglund-Liu [Liu (1976)] monodisperse atomizer (Model 3050), and droplet concentrations upstream of the shoot were measured by counting droplets deposited on a 1.0mm wide strip of Kromekote paper mounted on a thin wire across the test section. Shoot orientations with respect to the velocity vector were defined as:

a) the upper face of the shoot at a right angle to the flow ("normal"),
b) stem parallel, shoot facing upwards into the flow ("end"),
c) upper face at 45 deg ("oblique"), and
d) edgewise with edge normal to the flow ("edge", for the fir shoots only).

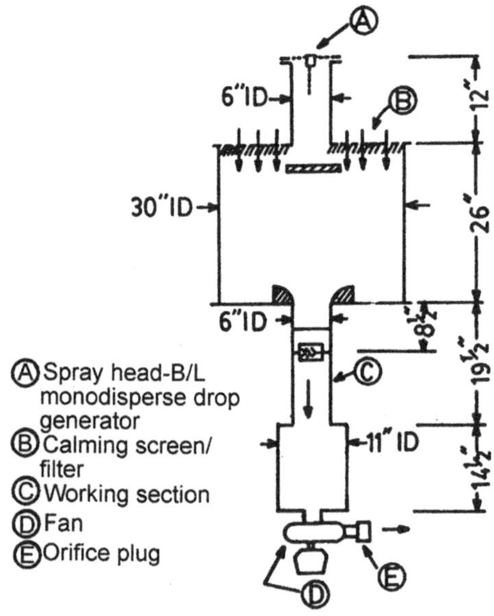

Figure 2.6. Shoot efficiency test section

Results were expressed using a needle collection efficiency as follows:

$$E_n = \frac{\text{(drops on needles per unit area per unit time)}}{\text{(upstream drop flux, drops/cm}^2\text{s)}} \qquad (2.3.3)$$

Because of the interaction between needles on a shoot, it is possible to determine a shelter factor, f, representing a reduction in the individual needle collection efficiency caused by a diversion of the flow streamlines around the shoot (a large obstacle to airflow, compared to an isolated needle). The use of the shelter factor is a first-order attempt to separate the inertial capture mechanics from the shoot geometry and external flow factors. Hence we expect f to be a function of velocity:

$$E_n = f.E_t \qquad (2.3.4)$$

Here E_t is the calculated May and Clifford ribbon collection efficiency for an isolated needle based on the approach velocity and mean needle width. Figure 2.7 shows values for E_n for normal orientation of fir shoots. The values of E_n obtained refer to the total foliage area of the needles in the shoot.Table 2.2 shows values of two constants a and b calculated from curve fits from the experiments assuming a linear relation between the shelter factor and upstream velocity:

$$f = a + b.V \qquad (2.3.5)$$

The values show a weak dependency on velocity, and significant scatter.

These values cannot describe the behaviour of the actual shoot-twig-branch-on-tree situation in collecting airborne particles because of the larger-scale interaction factors not represented in the laboratory experiments. The following section addresses this by describing the use and calibration of foliage simulators.

2.4 Foliage Simulators

Kristmanson et al. (1987) describe a foliage simulator (Figure 2.8) developed for the Dunphy trials (application of insecticide to coniferous species). This was done because of the need to have fresh collecting surfaces for droplets in successive experiments. (The continuing use of natural foliage during an experimental program creates severe problems because of the need to have "blank" uncontaminated foliage samples at each sample site before each test run. This is difficult in a multiple-test experimental program. Also, deposit analysis is simplified with simulators). The dimensions and shape of the simulator were

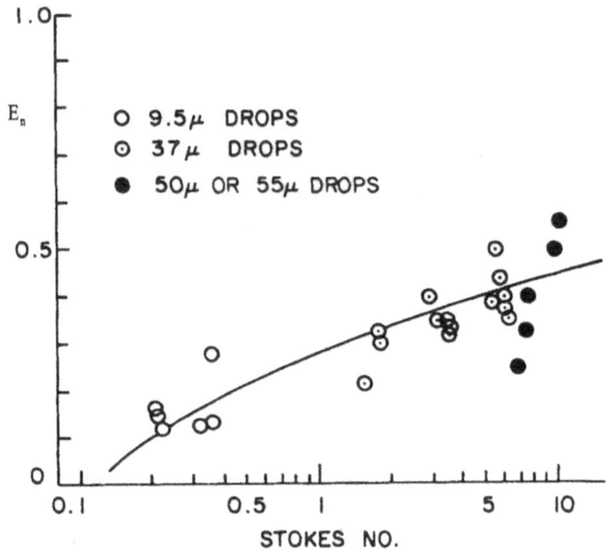

Figure 2.7. Envelope efficiencies for balsam fir shoots

TABLE 2.2. Shelter factor constants for balsam fir shoots

Species	Orientation	a	b	S.D.
Fir	Normal	.46	0	.073
	End	.213	.044	.025
	Edge	.095	0	.015
	Oblique	.290	.0804	.057
Spruce	Normal	.244	.1053	.046
	End	.140	.046	.042
	Oblique	.298	0	.065

Units of b are m^{-1} s

chosen to be aerodynamically similar to those of the shoots to be simulated, in this case, fir, spruce, and jackpine. Table 2.3 shows values of the shelter factor f found for this simulator in experiments identical to those described above for foliage shoots.

TABLE 2.3 Foliage simulator shelter factors (conifer)

Orientation to Velocity Vector	Constants for Shelter Factor*		
	a	b	r†
90°, VEE up	.647	-.040	.31
90°, VEE down	.392	.091	.57
45°, VEE up	.528	-.113	.81
45°, VEE down	.536	-.073	.57

*f = a + bV with V in m/s, see eq'n (2.3.5)
† correlation coefficient

NOTE: The calculated value for the 0° orientation is .04. The average of all orientations for an integral number of simulators is 0.34 at 1 m/s. Data not previously published.

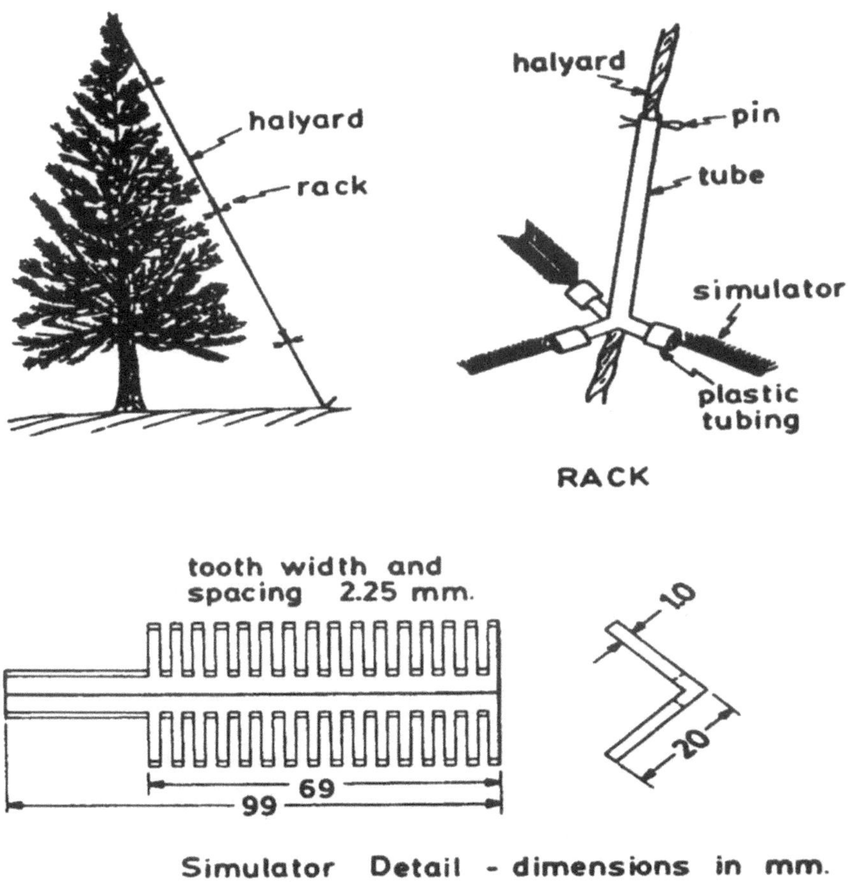

Figure 2.8. Simulator for conifer foliage, design and method of use

Figure 2.8 shows the deployment of simulators on lanyards strung on sample trees from the treetop down to the ground. The simulators were located at heights equal to the centroids of the top, middle, and bottom thirds of the cumulative leaf area distribution. It was found necessary to calibrate these simulators by comparing their total deposits against total deposits found on actual foliage samples adjacent to the simulators. The simulator lanyard kits were located so that upwind, downwind, and lateral positions on the tree were equally represented. The calibration procedure is discussed in Picot et al.(1993). The deposit of tracer or active ingredient on foliage is found by measuring deposit on a collection of simulators representing a sample location at a specific downwind distance. Typically, six simulators were used at each of the three vertical positions for each of three sample trees, located 30m apart in a direction parallel to the aircraft flight path. All these simulators were grouped together, washed with a chemical solvent to extract the tracer or active ingredient, and analyzed by fluorometry or by gas chromatography. Measurements gave the ratio of mass of tracer per gram of foliage to mass of tracer per simulator (C_i), directly, for each species i. This can be expressed as follows:

$$C_i = \{E_{f,i}/E_s\}.\ f_i / A_s \qquad (2.4.1)$$

Where $E_{f,i}$ = collection efficiency for actual foliage on tree, species i
E_s = collection efficiency for simulator
f_i = foliage area per gram for species i
A_s = area of one simulator (9.6 cm^2 for total flat cross-section of simulator tines).

The ratio $\{E_{f,i} / E_s\}$ was found from equation 2.4.1 when C_i was measured in calibration experiments with the appropriate value of f_i. Results are shown in Table 2.4. No significant correlation was found between this ratio and wind velocity for these tests. Data for the C_i determinations are given in van Vliet (1987).

When calibrated simulators are used in a field trial and simulator deposits are measured, foliage deposit can be calculated as follows:

$$\text{Fol. Dep.} = \text{Sim. Dep.} \times \Sigma_i \{(E_{f,i}/E_s)\ A_{f,i}\}/A_s \qquad (2.4.2)$$

Here, $A_{f,i}$ is the foliage area index for species i present in the stand, with units m^2 of foliage per m^2 of ground area. The simulator shelter factors of Table 2.3 are not involved in this calculation, but are useful in modelling calculations when used with the factors C_i or $E_{f,i} / E_s$ (see section **6.2** and Table 6.1).

Foliage simulators have been developed and used for broadleaf species as well [Picot and Kristmanson(1993)]. These are as shown in Figure 2.8 but are

TABLE 2.4 Simulator vs foliage calibration factors, C_i*.

	FIR	SPRUCE	JACK PINE
for tracer 1	.59 ± 0.034	.54 ± .034	.43 ± .022
for tracer 2	.75 ± .046	.49 ± .046	.49 ± .033
No. of Samples	29	29	29
Ave. C_i	0.67	0.52	0.46
cm^2/g**	15.3	15.5	18.6
$E_{f,i}/E_s$†	.42	.32	.24

* µg of tracer per g foliage/µg tracer per simulator, slope of the linear regression with standard error.

** needle area per unit mass (largest area in an axial plane) for fresh foliage not dried; area defined as in Table 2.1.

† ratio of droplets on foliage to droplets on simulators, for the same surface area.

made without notches so as to present solid surfaces to the airflow, with a total flat cross-sectional area of 27.6 cm^2. Values of the ratio $E_{f,i}$ / E_s were found by a procedure similar to that explained above for red maple and wild raspberry twig-leaf structures about 25 cm in length. The value for red maple was 0.70, and that for wild raspberry was 0.83.

2.5. Importance of Properly Defining the Biological Interface

For the control of forest pests by aerial application, the aerodynamic characteristics of the target are important. It is therefore essential to identify the target, or the "biological interface", on which active ingredient is to be deposited. In this chapter, the insecticide spraying referred to is designed to place a deposit on conifer needles for contact or ingestion by a mobile insect (larva) defoliator. The herbicide spraying referred to is designed to place a deposit of phytotoxic material on the surface of unwanted broadleaf species in a "conifer release" situation, for the elimination of broadleaf competition in a conifer plantation. In subsequent sections below, it will become clear that different droplet sizes should be used for foliage deposition and for ground deposition (using a systemic herbicide). When depositing material on foliage, the filtering effect of the foliage must be optimised. When depositing material on the ground, this filtering effect must be minimized.

References

Armstrong, J.A., and W.N. Yule (1978): The distribution of aerially-applied spray deposits in spruce trees, Can. Entomol. **110**, pp1259-1267.

Belot, Y., A. Baille, and J.-L. Delmas (1976): Modele numerique de dispersion des polluants atmospheriques en presence de couverts vegetaux, Atm. Env. **10**, pp 89-98.

Bergen, J. D. (1979): Aerial spray patterns in a forest canopy, Trans. ASAE, **22,** pp 1279-87.

Chapman, T.J. (1993): Modelling of Herbicide Spray Dispersion and Deposition, M.Sc. Thesis, UNB Dept. of Chemical Engineering, Fredericton, N.B.

Crabbe, R.S. , L. Elias, M. Krzymien, and S. Davie (1980): New Brunswick forestry spray operaions: Field study of effect of atmospheric stability on long range pesticide drift, National Research Council Lab Tech. Report #LTR-UA-52, Ottawa.

Davies, C.N., and C.V. Peetz (1956): Proc. Roy. Soc. (London), **A234**, 269.

Geiger, R. (1966): The Climate Near the Ground, Harvard University Press, Cambridge, Mass.

Grim, B., J. Rafferty, G. Sutton, and T. Clarke (1992): Deposition of Bacillus Thuringiensis into Gambel Oak Canopies, USDA Forest Service, Forest Pest Management, Davis , CA, #FPM 92-9.

Kasischke, E.S. (1995): Correlating radar backscatter with components of biomass in loblolly pine forests, IEEE Trans. on Geosc. and Rem. Sens. **33**, pp 643-659.

Ker, M.F. (1984): Biomass Equations for Seven Maritimes Tree Species, Info. Rept. M-X-148, Maritimes Forest Research Centre, Canadian Forestry Service, Fredericton, N.B.

Kristmanson, D.D., J.J.C. Picot, and G.W. Henderson (1984): The inertial impaction of droplets on foliage of some conifers, NRC Symp. on future role of aviation in agriculture, Saskatoon, Sask., April 1984, NRC Ref. AFA-TN-17, NRC 23504, Ottawa.

Kristmanson, D.D., J.J.C. Picot, S. Van Vliet, and G.W. Henderson (1987): Measuring foliar deposits from aerial spraying of pesticides, NRC Symp. on the aerial application of pesticides in forestry, Ottawa, October 1987, NRC Ref. AFA-TN-18, NRC 29197, Ottawa.

Langmuir, I., and K.B. Blodgett, (1944-5): General Electric Research Laboratory, Schenectady, N.Y., Report R.L. 225.

Liu, B.Y.H. (1976): Standardization and calibration of aerosol instruments, in: Fine Particles, B.Y.H. Liu ed., Symp. on fine particles, Minneapolis, 1975, Academic Press, N.Y.

May, K.R., and R. Clifford (1967): The impaction of aerosol particles on cylinders, spheres, ribbons, and discs, Ann. Occup. Hyg., **10**, pp 83-95.

Monteith, J.C. (1973): Principles of Environmental Physics, Edward Arnold Pub. Ltd., London, England.

Picot, J.J.C. (1986): Notes on vertical foliage distribution, unpublished, UNB Dept. of Chemical Engineering, Fredericton, N.B.

Picot, J.J.C., D.D. Kristmanson, R.E. Mickle, R.B.B. Dickison, C.M. Riley, C.J. Wiesner, (1993): Measurements of folial and ground deposits in forestry aerial spraying, Trans.ASAE, **36,** pp1013-1024.

Raynor, G.S. (1971): Wind and temperature structure in a coniferous forest and a contiguous field, Forest Sci. **17,** pp 351-363.

Riley, C.M., (1992): Continuation of Studies to Examine the Effects of Aircraft Application Parameters on Deposit and Drift of Forestry Herbicides, Report #C/92/443 (a), Research and Productivity Council, Fredericton, N.B., E3B6Z9

Rosenberg, N.J. (1974): Micro-Climate, the Biological Environment, Wiley/Interscience, J. Wiley & Sons, N.Y.

Seinfeld, J.H. (1975): Air Pollution, Physical and Chemical Fundamentals, McGraw-Hill, N.Y.

Steinke, W.E., T.M. Ledson, and N.V. O'Connell (1996): assessment of citrus canopy density and distribution with the Licor LAI-2000 leaf area index meter, paper #961080, 1996 ASAE Annual International Meeting, Phoenix, July 1996; ASAE, St. Joseph MI.

Strauss, W., (1975): Industrial Gas Cleaning, 2nd ed'n, Pergamon Press, Oxford, England.

Sun, G. And K.J. Ranson (1995): A three-dimensional radar backscatter model of forest canopies, IEEE Trans. on Geosc. and Rem. Sens. **33**, pp 372-382.

van Vliet, S.J. (1987): Aerial Spraying of Forests: Measurements of Ground and Canopy Deposits, B. Sc. Thesis, Dept. Of Chemical Engineering, UNB, Fredericton, N.B.

Wallace, D.J., J.J.C. Picot, and T.J. Chapman (1995): A numerical model for forestry aerial spraying, Agric. And For. Met. **76**, pp 19-40.

Wedding, J.B., T.C. Carney, R.B. Ekblad, M.E. Montgomery, and J.E. Cermak (1978): Particle deposition in a Douglas fir-canopy: A wind-tunnel simulation of aerial spraying, Trans. ASAE **21**, pp 253-260.

CHAPTER 3

PROPERTIES OF SPRAY FORMULATIONS

In a review on pesticide formulations, Sundaram (1995) has discussed the following factors important in their application: 1) storage, mixing, and pumping to the atomizing devices, 2) liquid atomization and droplet formation, 3) droplet transport and deposition, 4) droplet retention and spreading on target surfaces, and 5) deposit availability and mode of entry into insects. These factors are all taken into account in the preparation of formulations of the required toxicity to the target pest. Although items 4) and 5) are of primary importance, they are the province of the formulation chemist and will not be discussed herein, the assumption being that at the application stage, design problems should already have been solved. For further detail, the reader is directed to the ASTM (American Society for Testing and Materials) publications entitled "Pesticide Formulations and Application Systems" in the STP (Special Technical Publication) series. These contain papers presented at the Symposia of that name sponsored by ASTM Committee E35 (Pesticides) and E35.22 (Pesticide Formulations and Application Systems). The 14th volume appeared in 1995 [eds. F.R. Hall, P.D. Berger, and H.M. Collins (1995)] as ASTM STP 1234.

The effects of formulation viscosity, surface tension, and volatility on the first three items are discussed below. The initial section covers the problem of biological toxic requirement and its role in setting the droplet size and density for an effective deposit.

3.1. Biological Toxic Requirement and Droplet Size

As previously mentioned, normal practice in defoliation control is to apply pesticide to the foliage, where it will come into contact with the insect in the course of feeding and habitat-spinning activities. The accumulation of a lethal dose by the insect depends on the toxicity and concentration of the active ingredient in the droplets or granules deposited on the foliage, the droplet size, the number deposited per unit area, and the activity of the insect. Hall(1989) points out the complexity of the problem and the need to understand the reaction of the pest to the applied toxin, e.g. the effects of non-lethal doses which may cause insects to drop from trees, resulting in reduced defoliation.

Recent field trials on budworms in North America provide excellent insights. Kettela (1995) reports on an application of Bacillus thuringiensis (Berl.)

(B.t.) to a white spruce (Picea glauca (Moench) Voss) plantation in New Brunswick. The formulation (Foray 76B supplied by Novo Nordisk) was applied over a test site with a Cessna 188 aircraft using four Micronair AU4000 atomizers operating at 10,000rpm with a flow rate of 2 l/min per atomizer. Twelve passes over the same flight path were used per experiment on average, for five experiments. Figure 3.1 shows a comparison between the droplet spectrum generated by the aircraft, and the deposited spectrum on white spruce needles 50m downwind of the flight path. The generated spectrum was determined using the Malvern droplet spectrometer in the wind tunnel in the authors' laboratory [Murray (1991)] and is based on a laser beam transect through the atomizer plume centre-line (see Chapter 4 for details). The deposited spectrum was determined with a microscope by examining needles taken from the site and reporting droplet numbers in seven 25-micron size classes starting from for the overall range 1.0 to 175. microns. The generated spectrum $D_{v.1}$, $D_{v.5}$, and $D_{v.9}$ were 32, 55, and 95 micrometers respectively ($D_{v.5}$ is the drop size below which 50% of the volume is found, etc.). The number average droplet size measured on foliage was 64 micrometers, and on average, 93% of the droplets were less than 100 micrometers in diameter, and 70% of all droplets were in the 25 to 75 micrometer range.

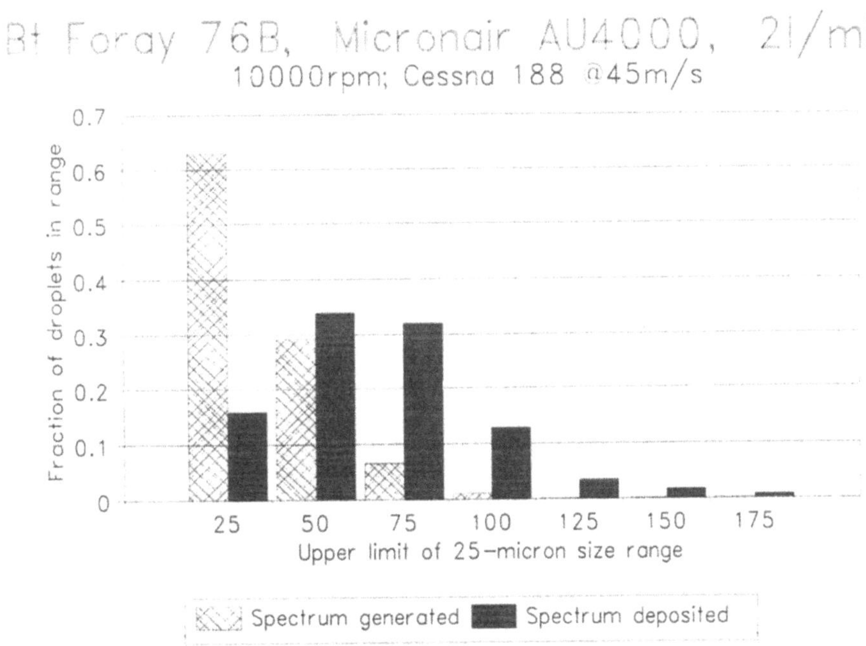

Figure 3.1 Droplets generated and deposited at 50 m downwind of spray line, white spruce plantation at 1600 stems/ha, 9.2 m mean tree height. Lower limit of drop size resolution is about 10 μm [Kettela (1996)].

The data clearly show the higher deposition efficiency of larger as compared to smaller droplets, but the sub-25 micrometer droplets are still present in significant numbers. Average droplet density for upper crown, mid crown, and lower third of crown at the 50m downwind location were 1.6, 1.1, and 0.2 droplets per needle, respectively. Average eastern spruce budworm survival was 0.25% for the 50m downwind location and 21% for the unsprayed control site, while defoliation was 6.5% in the treated block as compared to 28% in the unsprayed site. On average, there were 4.1 larvae per 45cm branch prior to spray application. A similar trial was conducted near Ste. Anne des Monts, Quebec, in 1990 [Kettela and Steel(1990)] with the same application system over an area heavily infested with eastern spruce budworm, 16.1 larvae per branch, in a maturing balsam fir forest. The B.t. formulation was Futura XLV-HP (Chemagro Ltd.) atomized to produce a similar deposit drop size distribution as indicated above. For a "normal application timing" spray, i.e. with the larvae at Instar IV or V, there was an almost linear correlation between reduction in defoliation and deposit per needle, 20% for 0.5 drops/needle, and 80% for 3.0 drops/needle. Defoliation in unsprayed control sites averaged 75%.

The above field research indicates that satisfactory protection against defoliation by the eastern spruce budworm can be achieved with a potent toxic active ingredient with a deposit density between 1 and 2 drops/ needle and a spectrum as shown on Figure 3.1. Similar results were recorded by Riley et al (1992) for a field trial of various formulations of B.t. applied against the western spruce budworm (Choristoneura occidentalis, Freeman) in a Douglas fir and grand fir forest in the Umatilla National Forest in Oregon. In that study, the deposit $D_{v.5}$ and $D_{n.5}$ were 106 and 52 micrometers, on average, due to the use of different equipment.

The selection of proper dosage levels appears to require field trials of the type mentioned above, even when laboratory data on dose response is available. This is illustrated by a comparison between laboratory and fields studies on the growth regulator RH5992 (Mimic 2F, from Rohm and Haas), which is discussed in detail in section 8.3 below.

Similar complications are found in the use of herbicides. For instance, in a laboratory study of the effects of adjuvants and droplet size on the efficacy of a glyphosate herbicide, Prasad (1992) found that droplets in the range 159 to 332 micrometers in diameter were generally more phytotoxic to white birch (Betula papyrifera L.) than 575 micrometer droplets, for the same total mass of glyphosate applied.

The data quoted above indicate that in a situation where adequate application methods are in use, a more detailed knowledge of the biological requirement for pest control may permit modifications resulting in improved control and reduced pesticide requirement. This will be addressed further in Chapter 8.

3.2 Effects of Viscosity, Surface Tension, and Volatility

Viscosity

Pesticide formulations can be purely viscous (Newtonian), or non-Newtonian in the sense of viscosity being dependent on shear rate in the fluid. They can also exhibit a degree of viscoelasticity when high molecular weight polymer additives are added in an attempt to reduce the production of small droplets in the atomizer (these are usually called drift-reducing additives). The viscosity of the formulation has an effect on the achievable flow rate in a pesticide delivery system, but experience has shown that it has a minor influence on drop size distribution from atomizers used in aerial spraying, for a given flow rate.

The prediction of pressure drop for a given flow rate of a Newtonian fluid is described in many engineering fluid mechanics textbooks [e.g. Daugherty et al. (1985)] and is sufficiently well known that it need not be repeated here. For non-Newtonian fluids the methods are less well known and are briefly described below, for cases where flow is steady-state from the point of view of a moving reference frame in the fluid (i.e. steady flow in constant diameter pipes downstream of the entrance region).

The flowrate for a given pressure drop across a piping system for time-independent flows can be estimated using the method recommended by Dodge and Metzner (1959). The basis of the method is the non-Newtonian power-law model for shear stress τ as a function of shear rate du/dr when velocity u depends solely on position r:

$$\tau = K\left|\frac{du}{dr}\right|^{n-1} \cdot \frac{du}{dr} \qquad (3.2.2)$$

Here K is the viscosity constant and n is the power-law constant. This expression fits a wide range of non-Newtonian fluids, and the Newtonian case is satisfied with $n = 1.0$ and $K = \mu$ (absolute viscosity). Pressure drop in pipe flow due to viscous friction is found by computing the friction factor f using a modified Reynolds number expression in equation 3.2.3 below:

$$Re' = \frac{D^n V^{2-n} \rho}{g_c K \left(\frac{3n+1}{4n}\right)^n} 8^{n-1} \qquad (3.2.3)$$

Here, D is pipe internal diameter, V is pipe average flow velocity, and g_c is the mass-weight conversion factor, 32.2 ft.lbs-mass per lb-force s^2 (Engineering units) or 1.0 kg m per Newton s^2 (S.I. units). Fluid density is ρ. The dimensionless pressure drop is the Fanning friction factor defined as:

$$f = \frac{D\Delta P}{2\ L\ \rho\ V^2} \tag{3.2.4}$$

ΔP is the pressure drop for a smooth-walled pipe or tube of length L. For laminar flow (below $Re' = 2100$), the relationship between friction factor and the Reynolds number is:

$$f = \frac{16}{Re'} \tag{3.2.5}$$

For turbulent flow (above $Re' > 3100$), the appropriate relation from Dodge and Metzner (1959) is:

$$\sqrt{\frac{1}{f}} = 4.0n^{-0.75}\log\left(Re'\ f^{1-\frac{n}{2}}\right) - \frac{0.40}{n^{1.2}} \tag{3.2.6}$$

In the example that follows, the relation between flow rate and pressure loss is estimated for a typical formulation of a bacterial agent. The pressure losses in straight pipe (say, a boom) will be compared for the cases of water and the formulation.

EXAMPLE

In using the above equations to calculate a pressure drop for a given flow rate of fluid in a pipe of specified diameter and length, it is necessary to determine the constants K and n from measured values of viscosity at a rage of shear rates.

Data for Bt Foray 76B measured with a Haake Rotovisko viscometer are given in the table below:

TABLE 3.1 Measured viscosities at calculated surface shear rates, Foray 76B B.t. at 20°C

Visc., Pa.s	Shear rate, s^{-1}
0.513	73.1
0.363	146
0.271	218
0.185	435

Note:
1. Water viscosity at this temperature is 0.001 Pa.s (or kg/m s)
2. A linear regression curve fit yields K=6.275, and n = 0.401

Using the values of Table 3.1, the pressure drop for a 10 l/min flow through a 2.54cm diameter smooth tube per meter of length can be calculated for both water and Foray 76B:

<u>Water:</u>

$V = 10{,}000/(60\ \pi D^2/4) = 167 \times 4/(\pi \times 2.54^2) = 33$ cm/s

$Re = 0.33$ m/s $\times$ .0254 m $\times$ 1,000 kg/m^3/.001 Pa.s

$= 8355$ (turbulent)

from eq'n 3.2.6: (n=1.0)

$$(1/f)^{0.5} = 4.0 \times \log(8355 \times f^{.5}) - 0.4$$
$$f = 0.0081$$

then $\Delta P = f/D \times 2\ L\ \rho\ V^2 = .0081/.0254 \times 2 \times 1.0 \times 1000 \times .33^2$

$= 35$ Pa per m of pipe length (0.005 psi per m)

<u>Foray 76B:</u>

$V = 0.33$ m/s

$$Re' = .33^{(2-.40)} \times .0254^{.40} 1{,}070/(6.275((3 \times .40+1)/(4 \times .40))^{.40} \times 8^{(.1-.40)})$$
$$= 61 \text{ (laminar)}$$

then $f = 16/61 = 0.26$

and $\Delta P = .26/.0254 \times 2 \times 1.0 \times 1070 \times .33^2 = 2386$ Pa per m (0.35 psi/m)

The calculation shows that for a 10m run of the 2.54 cm diameter (1.0 inch) pipe, the pressure drop for the Foray 76B is only 3.5 psi. Even though it is 70 times greater than the pressure drop using water, it will not have a very significant effect for most aircraft spray delivery systems in forestry use. However, the calculations outlined above are necessary to permit a preliminary estimate of pressure drop in the pesticide delivery system leading up to the atomizers, for pesticides having significantly higher viscosity. Viscometer shear rates used should be in the same order of magnitude as the value at the pipe wall, which can be estimated to be approximately equivalent to 2(V/D). A standard text in fluid mechanics [Daugherty et al. (1985)] can be consulted for pressure drop through fittings, and it may be necessary to do laboratory tests to get pressure drop through nozzles.

Viscosity has a minor effect on atomizer performance for the atomizers used in forestry (rotating cages, flat fan, and other types of one-fluid nozzles using the aircraft slipstream for atomization). Picot et al (1989) have shown that the $D_{v.5}/d_o$ ratio of nozzles correlates with the dimensionless group $Re^{-.059}\ We^{-.755}$ (for viscosities from 2 to 260 mPa.s) where d_o is the characteristic diameter of the final orifice ejecting fluid into the air stream (screen mesh effective pore size, for example, or film thickness for a spinning disc atomizer), and Re is $d_o V\rho/\mu$ with V being relative velocity between fluid and air at the point of exit from the atomizer, and ρ and μ being the density and absolute viscosity of the fluid, respectively. The Weber number We used in the above work is defined as $V^2\rho d_o/\sigma$ where ρ is fluid density and σ is fluid-to-air surface tension (not the standard definition where the density term is for air [Lefebvre (1989)]). This means that for two fluids of different viscosities when other properties and operating factors are identical, the effect of viscosity on $D_{v.5}$ is equivalent to the ratio $(\mu_2/\mu_1)^{.059}$. This means a 1000-fold increase in viscosity is required to cause a 50% increase in $D_{v.5}$. Akesson and Gibbs (1990) also found very little effect of viscosity on the $D_{v.5}$ of the drop spectrum. They also show the effect of increasing concentration of soluble polymers on drop size for both a cone and a fan nozzle. For a range of polymer concentration from 26 to 900 ppm, the $D_{v.1}$ increased slightly, while the $D_{v.5}$ increased 100%. Chapple et al (1992) show that for cases where a significant increase in $D_{v.5}$ was achieved with a "drift-control" additive, the increase was attenuated by recirculating through pumps, the soluble polymer additive apparently being degraded by fluid shear. General conclusions are impossible, as the behaviour depends on the elasticity and extensional viscosity and is very complex [Bousfield et al (1986)]. Wind tunnel tests are needed for specific cases.

Surface Tension

Akesson and Gibbs (1990) point out that surface tension has a significant effect on drop size, but that most pesticide formulations have surface tension values in the range 35 to 40 mN/m when in a water carrier. The Picot et al correlation quoted above indicates that for other factors being equal, the effect on the $D_{v.5}$ ratio for two liquids of different surface tension is the factor $(\sigma_1/\sigma_2)^{-.755}$ i.e. lower values of surface tension σ yield smaller droplets. On the other hand, Elliott and Wilson (1983) state that the exponent of this ratio lies between -0.50 and -0.67. Because of the narrow range of surface tension values available for most studies (water has a surface tension of 72 mN/m), accurate dependency on this factor alone is difficult to measure. An estimate of the $D_{v.5}$ for atomization of a fluid in an atomizer when previous measurements with another fluid are available can be made by using the value -0.7.

Lefebvre (1989) presents a comprehensive analysis of the performance of nozzles used in combustion systems, giving correlations for mean diameters for each type, but data for atomizers used in forestry or agriculture are limited.

Volatility

The evaporation of droplets is important in that it affects the settling and deposition characteristics of the drop on target surfaces. In most cases, the carrier liquid or diluent is the only component undergoing significant evaporation, the active ingredient being of low volatility. In general, the Goering model [Goering et al (1972)] can be used to calculate the change in droplet diameter when the drop is moving relative to ambient air:

$$\frac{-\delta D_t}{\delta t} \cdot D_t \left[2.0 + 0.6\ R_e^{0.5}\ S_c^{0.33}\right]^{-1} = \frac{2.0\ M_v D_v \rho_a \Delta p}{(M_m \rho_l P_f)} \qquad (3.2.7)$$

where

D_v	=	Molecular diffusivity of evaporating vapour in air, m^2/s
D_t	=	Mean drop diameter between time t and t+δt, m
M_v	=	Molecular weight of evaporating vapour, kg
M_m	=	Mean molecular weight of resulting air-vapour mixture, kg
Δp	=	Partial pressure of the evaporating liquid at the drop surface minus the partial pressure of that component in the ambient air stream, Pa
P_f	=	Partial pressure of air, Pa
Re	=	Reynolds number, $D_t\ V\ \rho_a / \mu$, dimensionless
Sc	=	Schmidt number, $\mu / \rho_a D_v$, dimensionless
t	=	Elapsed time from drop generation, s
V	=	Relative velocity between drop and air stream, m/s
ρ_a	=	Density of air at mean temperature between drop surface and ambient air (film temperature), kg/m^3
ρ_l	=	Density of liquid, kg/m^3
μ	=	Viscosity of air at film temperature, Pa.s

The term $\delta D_t / \delta t$ is the rate of change in the drop diameter with time. The Reynolds number includes the relative velocity between drop and air, which in this case is the terminal settling velocity of the droplet in air. Equation 3.2.7 reduces to the theoretical solution at zero velocity when the term in square brackets becomes equal to two. Wedding et al (1986) present a correlation based on evaporating water droplets in simulated free-fall conditions and give the constants 1.755 and 0.535, instead of 2.0 and 0.6 for the term in square brackets. Riley et al(1995) show that equation 3.2.7 can be numerically integrated to give diameter after time of generation, using short time steps to permit the introduction of up-dated drop size and terminal settling velocity as time progresses, for liquids where vapor pressure is known and the temperature of the drop can be accurately calculated using a heat balance for the drop with a heat transfer coefficient calculated from the heat transfer analog of equation 3.2.7. [Perry and Chilton (1973)]. Liquid vapor pressures (when discussing the gas phase these become partial pressures) for ideal solutions can be found by using Raoult's law, but this is of marginal use for pesticides because of the complex nature of most formulations. Holder (1979) found that acceptable accuracy was obtained for a water-based emulsion when the properties of water were used in equation 3.2.7, down to the point of breakdown of the emulsion, when most of the water had evaporated. Figure 3.2 shows experimental results from this work as well as calculated results for a range of water droplet sizes.

Picot et al (1981) developed a method for dealing with multi-component solutions or mixtures based on the observation that for many pesticides, droplets were well-mixed due to internal circulation. Vapor pressure depends on drop composition (and temperature), and composition is gradually changing with the more volatile components evaporating more quickly. Properties therefore can be expressed as a function of the ratio of the diameter at any time to the initial diameter. In this case, the complete right-hand side of equation can be replaced by a polynomial expressing all of the properties in that term, for cases when the partial pressure of the evaporating material in the ambient air is negligible:

$$\frac{-\delta D_t}{\delta t} \cdot D_t \, [\, 2.0 + 0.6 \, Re^{0.5} \, Sc^{0.33}]^{-1} = a + b\left(\frac{D_t}{D_i}\right) + c\left(\frac{D_t}{D_i}\right)^2 \qquad (3.2.8)$$

The constants a, b, and c must be determined experimentally. The Picot et al (1981) experiments were on a petroleum oil and the experimental set-up was simple, involving a measured air flow, visual observations through a microscope and a stopwatch, with droplets suspended on a fine wire. The Riley et al (1995) experimental set-up (see Figure 3.3) is more sophisticated, involving video-computer data acquisition and automatic control of temperature and relative humidity, required for a study of water-based solutions.

The Picot et al (1981) experimental data on droplets of a #2 fuel oil with most of the cyclic compounds removed and 5% of tracer (2-ethyl hexyl phosphate, essentially non-volatile under these conditions) added in solution are shown in Figure 3.4. Curve fit equations for the two temperatures studied for a range of sizes are given below the Figure. These equations are useful for drop dispersion and deposition models where trajectory calculations are up-dated every fraction of a second (see Chapter 6). Although theoretical calculations are possible for multi-component drop evaporation, these would still require basic vapor pressure data for complex mixtures, requiring extensive laboratory measurements. The assumption of well-mixedness can be tested by studying different sized droplets which should give the same polynomial fit at a given temperature. The simple experiments described here are therefore recommended.

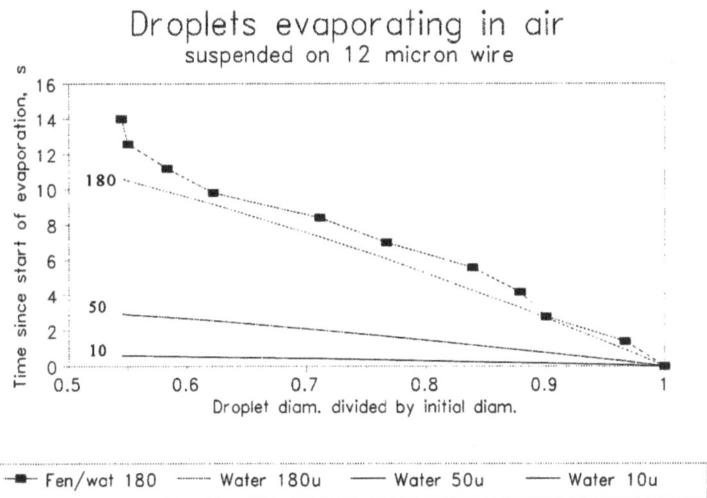

Figure 3.2 Comparison of experimental data for evaporation of a water-based fenitrothion emulsion droplet, and theoretical values from equation 3.2.8 for pure water, for droplet initial diameters of 180, 50, and 10 micrometers. Air flow past the droplets is 0.88 m/s which is slightly higher than the terminal settling velocity of the 188 micrometer droplet but accounts for only 8% of the rate of diameter reduction at 180 microns, and about 2.5% at 10 microns [Holder (1979) data for 180 micrometer droplets]. Temperature 21.7° C, relative humidity 18%. The emulsion contains 9.6% fenitrothion, 2.7% solvent (Dowanol), 2.7% emulsifier (Atlox), and 85% water (all vol%). The evaporation of the emulsion follows the theoretical curve for water until most of the water has evaporated.

The second-order curve fits obtained by regression analysis at the two temperatures 21.4° C (A:) and 4.7°C (B:) are as follows:

A: for $0.5 < \dfrac{D_t}{D_i} < 1.0$:

$$D_t\left(\frac{-\delta D_t}{\delta t}\right)\left[2.0 + 0.6\ Re^{0.5} Sc^{0.33}\right]^{-1}$$

$$= \left[14.88 - 57.5\left(\frac{D_t}{D_i}\right) + 57.5\left(\frac{D_t}{D_i}\right)^2\right] \times 10^{-12}$$

and for $0.45 < \dfrac{D_t}{D_i} < 0.5$:

$$= \left[-4.5 + 10.0 \frac{D_t}{D_i} \right] \times 10^{-12} \tag{3.2.9}$$

B: for $0.65 < \frac{D_t}{D_i} < 1.0$:

$$= \left[30.5 - 92.33\left(\frac{D_t}{D_i}\right) + 71.0\left(\frac{D_t}{D_i}\right)^2 \right] \times 10^{-12}$$

and for $0.45 < \frac{D_t}{D_i} < 0.65$

$$= \left[-0.70 + 1.56\left(\frac{D_t}{D_i}\right) \right] \times 10^{-12} \tag{3.2.10}$$

The units of all equations are m^2/s. As mentioned above, a different correlation will result for each fluid and temperature.

Herbicide formulations are often dilute solutions of active ingredients in water. The Δp term in equation 3.2.7 will then require a Raoult's law calculation to get the vapor pressure of water at the drop surface:

$$\Delta p = P_{w,s} - P_{w,a} = \alpha \cdot x \cdot P_w^o - P_{w,a} \tag{3.2.11}$$

Here, x is mole fraction of water in the solution, α is the activity coefficient for water in the solution, and P_w^o and $P_{w,a}$ are the vapor pressure of pure water at the temperature of the droplet and the partial pressure of water vapor in the ambient air.The value of α is close to unity for dilute solutions, but can deviate significantly from unity as the concentration of the solute increases as the droplet evaporates, since the active ingredients usually have much lower vapor pressures than water. The value of α can be calculated from the experiments mentioned above as droplet evaporation proceeds, and can be expressed as a polynomial:

$$\alpha = d + e\left(\frac{D_t}{D_i}\right) + f\left(\frac{D_t}{D_i}\right)^2 + g\left(\frac{D_t}{D_i}\right)^3 \tag{3.2.12}$$

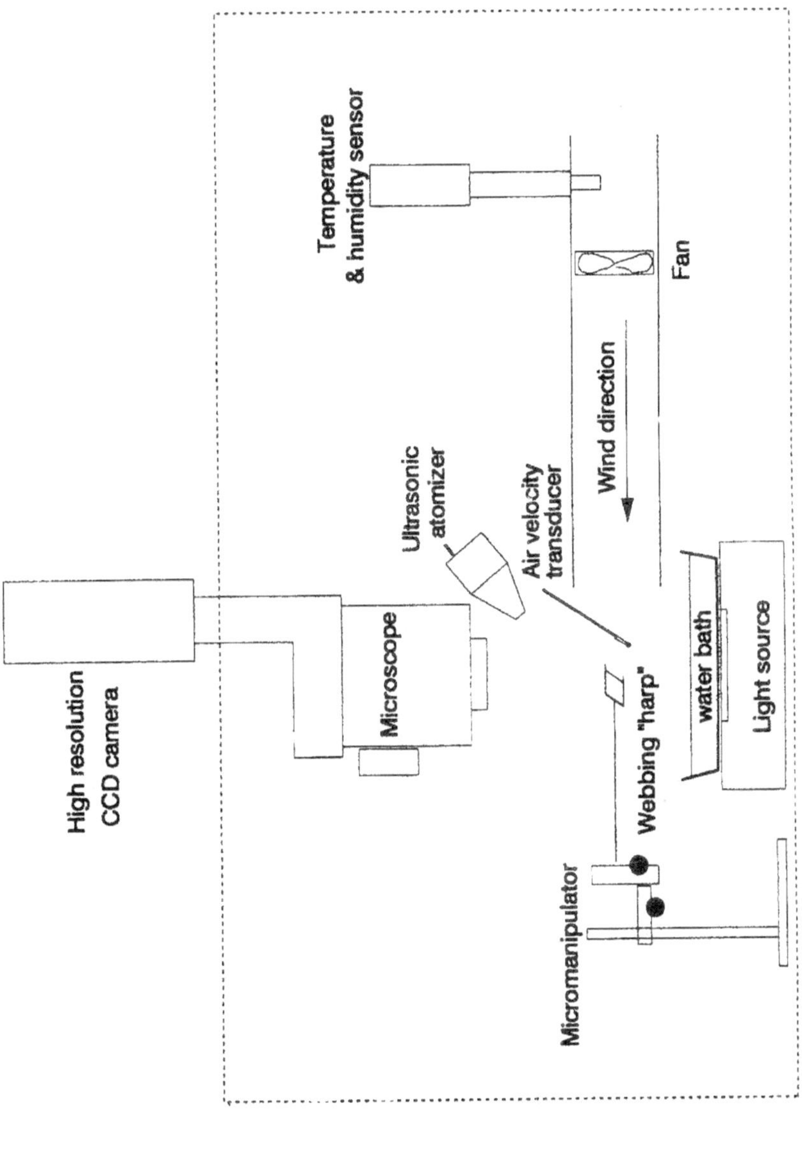

Figure 3.3 Apparatus for drop evaporation measurements, copied from Riley et al (1995) with the kind permission of the American Society for Testing and Materials.

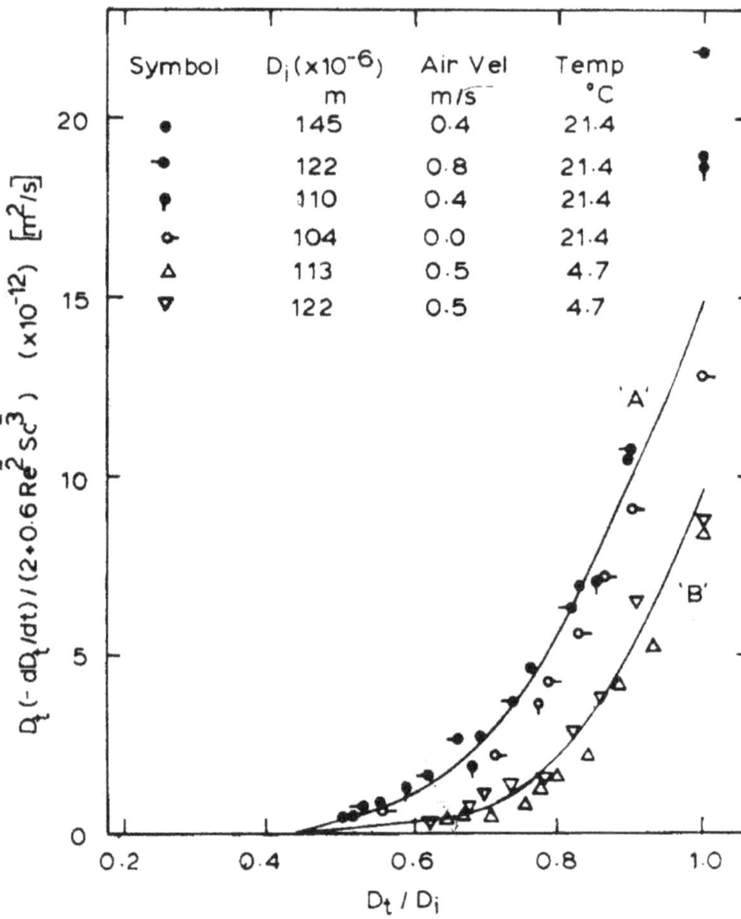

Figure 3.4 Experimental data and polynomial curve-fits for evaporation of #2 fuel oil (see text) droplets, copied from Picot et al (1981) with the kind permission of the American Society of Agricultural Engineers.

The constants in 3.2.12 are found by regression analysis on the series of α values found from evaporation measurements as time progresses. The value of P_w° is the equilibrium vapor pressure for pure water at the droplet surface, which is cooler than the surroundings and calculated from wet and dry bulb measurements of the ambient air, based on data given in humidity charts [Himmelblau (1982)]. Once the equation 3.2.12 constants are known, a step-wise numerical integration of equation 3.2.8 gives droplet diameter for any time after droplet generation. In general, there will be a specific equation for each formulation at each temperature.

3.3 Summary

A pesticide delivery system must be carefully designed to minimize the application rate, for both economic and environmental reasons. A detailed field study of pesticide dose response has the potential to improve pest control with reduced application volume, by more accurately defining the dosage rates for pest control. Because of the wide variation in formulation properties, atomizer types, and aircraft performance parameters, it is difficult to understand the mechanics of spray generation and deposition without recourse to experiments and calculation. For very high viscosity liquid formulations (including solid suspensions in liquids), laboratory tests should be carried out to check flow rates in delivery equipment and atomizers. Wind tunnel atomizer characterization is essential with new atomizer types and formulations having novel flow properties. With complex formulations, experimental studies on droplet evaporation may be required to assess the change in the droplet spectrum during the dispersion and deposition process.

References

Akesson, N.B., and R.E. Gibbs (1990): Pesticide droplet size as a function of spray atomizers and liquid formulations, in Pesticide Formulations and Application Systems, volume 10, L.E. Bode, J.L. Hazen, and D.G. Chasin eds., STP 1078, ASTM, Philadelphia PA, pp 170-183.

Bousfield, D.W., R. Keunings, G. Marucci, and M.M. Denn (1986): Nonlinear analysis of the surface tension driven breakup of viscoelastic filaments, J. Non-Newt. Fl. Mech. **21**, pp 79-97.

Chapple, A.C., R.A. Downer, and F.R. Hall (1992): Effects of pump type on atomization of spray formulations, in Pesticide Formulations and Application Sysytems, volume 11, L.E. Bode and D.G. Chasin eds., STP 1112, ASTM, Philadelphia, PA, pp 193-205.

Dodge, D.W. , and A.B. Metzner (1959): Turbulent flow of non-Newtonian systems, A.I.Ch.E. J., **5**, pp 189-204.

Daugherty, R.L., J.B. Franzini, and E.J. Finnemore (1985): Fluid Mechanics with Engineering Applications, 8th ed'n, McGraw-Hill, New York, NY.

Elliott, J.G., and B.J. Wilson (1983): The Influence of Weather on the Efficiency and Safety of Pesticide Applications: The Drift of Herbicides, Occasional Public. #3, Report of a Working Party of The Research and Development Committee, The British Crop Protection Council, BCPC Publications, Croydon, England.

Goering, C.E., L.E. Bode, and M.R. Gebhardt (1972): Mathematical modelling of spray droplet deceleration and evaporation, Trans. ASAE, **15**, pp 220-225.

Hall, F.R. (1989): Effect of formulation, droplet size, and spatial distribution on dose transfer of pesticides, in Pesticide Formulations and Application Systems, 8th volume, D.A. Hovde and G.B. Beestman eds., STP 980; ASTM, Philadelphia, PA., pp 149-154.

Hall, F.R., P.D. Berger, and H.M. Collins eds. (1995): Pesticide Formulations and Application Systems, 14th volume; STP 1234; ASTM, Philadelphia, PA.

Himmelblau, D. M. (1982): Basic Principles and Calculations in Chemical Engineering, Prentice-Hall, Englewood Cliffs, NJ, p 485.

Holder, J.A. (1979): Evaporation Rates of Aerial Spray Droplets, B.Sc. Thesis, Dept. of Chemical Engineering, Univ. Of New Brunswick, Fredericton, N.B.

Kettela, E.G. (1996): personal communication, July 23, 1996.

Kettela, E.G. (1995): Evaluation of quality and quantity of B.t. spray deposit on white spruce. Research report to SERG (Spray Efficacy Research Group),Fredericton, N.B., available from Canadian Institute for Scientific and Technical Information (CISTI), Ottawa.

Kettela, E.G. (1993): Evaluation of RH5992 (Mimic 2F) aerial spray trials against spruce budworm in New Brunswick in 1992. Research report to SERG, Fredericton.N.B., available from CISTI, Ottawa.

Kettela, E.G. and V. Steel (1990): An account of spray trials conducted to evaluate the efficacy of B.t. against high spruce budworm populations. 1990. Research report to SERG, Fredericton, N.B., available from CISTI, Ottawa.

Lefebvre, A.H. (1989): Atomization and Sprays, Hemisphere Pub. Corp., New York, NY.

Murray, D. (1991): An analysis of B.t. sprays and simulants, Research report, Dept. of Chemical Engineering, Univ. Of New Brunswick, Fredericton, N.B.

Perry, R.H., and C.H. Chilton (1973): Chemical Engineers' Handbook, 5th ed'n, McGraw- Hill, New York, NY, p 18-51.

Picot, J.J.C., B. Chitrangad, and G. Henderson (1981): Evaporation rate correlation for atomized droplets, Trans. ASAE, **24**, pp 552-554.

Picot, J.J.C., M.W. van Vliet, and N.J. Payne (1989): Droplet size characteristics for insecticide and herbicide atomizers, Can. J. of Chem. Eng. **67**, pp 752-761.

Prasad, R. (1992): Some factors affecting herbicidal activity of glyphosate in relation to adjuvants and droplet size, in Pesticide Formulations and Application Systems, volume 11, L.E. Bode and D.G. Chasin eds., STP 1112, ASTM, Philadelphia PA, pp 247-257.

Riley, C.M., I.I. Sears, J.J.C. Picot, and T.J. Chapman (1995): Description and validation of a test system to investigate the evaporation of spray droplets, in Pesticide Formulations and Application Systems, volume 14, F.R. Hall, P.D. Berger and H.M. Collins eds., STP 1234, ASTM, Philadelphia, PA, pp 225-236.

Riley, C.M., C.J. Wiesner, D.W. Scott, J. Weatherby, and R.G. Downer (1992): Evaluating the field efficacy of B.t. Berliner against the western spruce budworm, in Pesticide Formulations and Application Systems, volume 11, L.E. Bode and D.G. Chasin eds. , STP 1112, ASTM, Philadelphia PA, pp 271-290.

Sundaram, A. (1995): Pesticide Formulations, Chapter 46 of Forest Insect Pests in Canada, J.A. Armstrong and W.G.H. Ives eds., Natural Resources Canada, Canadian Forest Service, Science and Sustainable Development Directorate, Ottawa.

Wedding, J.B., Y.J. Kim, and R.S. Dennison (1986): Mass transfer from water droplets under simulated free-fall conditions, Atm. Env. **20**, pp 1039-1045.

CHAPTER 4

SPRAY AIRCRAFT AND ATOMIZERS

4.1 Aircraft in Use

Various aircraft types have been used in aerial spraying [Quantick(1985a)]. These include the DC-6, the Grumman Avenger (TBM), the Dromader M-18, the CMELAK Z-37 T, and the the Bell 206 helicopter, all shown on Figure 4.1. Table 4.1 gives details of type, size, weight, payload, of spray aircraft registered in Canada in 1996.

4.1 a) DC-6 spraying spruce budworm, in Northern New Brunswick, 1988.
Photo courtesy of C. Dixon, Forest Protection, Ltd., Fredericton, N.B.

Figure 4.1 Examples of spray aircraft

4.1b) Grumman TBM ground testing SERG/ONAIR atomizers with water, Sevogle airstrip, New Brunswick

4.1c) Dromader M-18 equipped with SERG/ONAIR atomizer prototypes (see text below)

Figure 4.1(cont.) Examples of spray aircraft

4.1d) CMELAK Z-37 agricultural aircraft, SLOV-AIR, Bratislava, Slovak Republic

4.1e) Bell 206 helicopter

Figure 4.1(cont.) Examples of spray aircraft

TABLE 4.1 Spray aircraft registered in Canada, 1996

TYPE	NO	GROSS WEIGHT kg	PAYLOAD litres	WING SPAN metres
PIPER PAWNEE	81	1318	455	11
CESSNA 188	86	1500-2090	380-550	12.8
GRUMMAN AgCat A+B	52	2761-3000	682-872	11-14
AYRES THRUSH	31	3273	1137	14.6
AYRES BULL THRUSH	2	4136	1516	14.6
AYRES TURBO THRUSH	3	4136	1327	14.6
AIR TRACTOR 300-301	4	2273	1327	13.7
AIR TRACTOR 401-401 B	8	3573	1478	14.9
AIR TRACTOR 502 B	3	4273	1857	15.5
PZL M-18	20	5318	2198	17.7
GRUMMAN TBM	12	8173	2369	16.5

Data Compiled by C. Dixon, Forest Protection Ltd., Fredericton, NB

The selection, operation, management, and maintenance of spray aircraft are critical issues in any forestry spray program. The excellent books by Quantick (1985a, 1985b) cover these topics. The complex factors which come into play are beyond the scope of this book. We concentrate below on the characteristics of the aircraft which affect the generation of droplets and their delivery to the forest foliage, and the choice and characteristics of atomizers.

4.2 Aircraft Wake Behaviour

The aircraft wake downwash has an important effect on generated drops in that it gives an initial downward velocity to the droplet cloud. For an aircraft flying at constant altitude, the air flowing over the upper convex wing surface first acquires an upward velocity component as it moves up to the thickest part of the wing, then a downward velocity component as it moves to the trailing edge of the wing and beyond. This downward flow immediately behind the wing creates a discontinuity at both wing tips where it meets air which has no downward motion. This forms a vortex at each wing tip. When the aircraft starts its motion through the air, a vortex is formed downwind of the trailing edge, again due to flow discontinuity. This vortex parallel to the trailing edge is integral with the wing tip vortices and forms a complete three-dimensional vortex with the flow immediately about the wing. In ideal fluid flow conditions (zero viscosity), the parallel vortex stays behind at the airport but is linked to the wing through the ever-lengthening wing-tip vortices which have grown to a rolled-up vortex cell (Figure 4.2) a short distance behind the aircraft. This cell moves downward through the air since its downward momentum is equivalent to the weight of the aircraft.

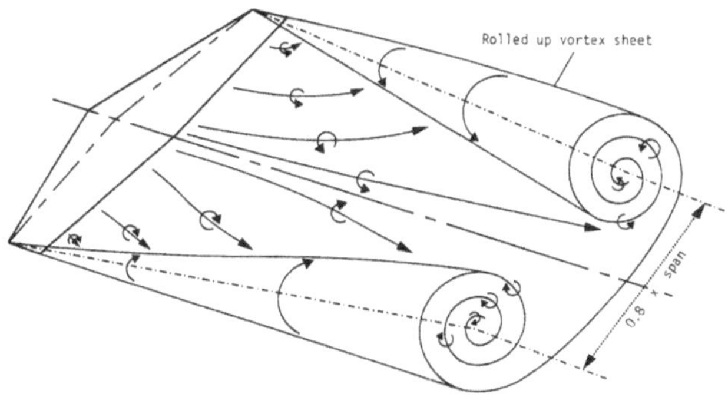

Figure 4.2 Vortex roll-up behind moving wing [Elliott and Wilson (1983)]. This grows to incorporate all of the trailing-edge air at a short distance behind the aircraft.

Hallock and Eberle (1977) have summarized the results of theoretical and experimental studies of aircraft wake vortex behaviour. Assuming ideal fluid behaviour for air, and an elliptic distribution of wing loading from tip to tip, the downward translational velocity V of the rolled-up (developed) vortex cell is:

$$V = \frac{8L}{(\pi^3 \rho U_\omega b^2)} \tag{4.2.1}$$

Here ρ is air density, U_ω is aircraft speed, b is wingspan, and L is aircraft weight. It can be seen that high downward velocity is favored by high weight, low air speed, and short, wide wings. Tennankore et al(1980) showed that for measurements on a Grumman (TBM) Avenger while spraying, complete vortex roll-up occurs 6 to 10 wing-spans to the rear of the aircraft and that vortex downward velocities were close to the value given by eq'n 4.2.1 for a time period of up to eight seconds after vortex initiation, after which time the vortex started to break up, apparently as a result of interaction with the atmosphere and the canopy. Brashears and Hallock (1974) describe a wake transport model which incorporates wind shear, buoyancy, flow induction, ground plane interaction, viscous decay, and core instability effects. Good agreement is shown with measurements on large aircraft. Wallace (1988) reviewed the status of theoretical simulation of aircraft vortex behaviour and chose the VORTEX WAKE code developed by the Aeronautical Research Associates of Princeton (ARAP) as giving realistic results for vortex rebound from a ground plane, and the more rapid dissipation of the downwind vortex in the presence of a crosswind. This model simulates the viscous and turbulent dissipation of wake-generated vortices immersed in a turbulent atmosphere [Bilanin et al (1977, 1978)]. Donaldson's (1973) second-order closure model is used to solve the Reynolds stress equations for a two-dimensional wake. The calculation is initiated by centering two counter-rotating, fully rolled-up vortices at distances of (+/-) $b\pi/8$ from the wing centreline. The vortex cores at these locations are treated as rotating like solid bodies and are given the following maximum vorticity for the elliptical wing loading case:

$$\eta_{max} = 4L \frac{g}{(r_c^2 \pi^2 \rho U_\omega b)} \tag{4.2.2}$$

Here g is the gravitational constant and r_c the vortex core radius. Vorticity is defined as twice the rate of angular rotation or "spin" in a flow and can be measured by a four-bladed vane immersed in the fluid.

VORTEX WAKE requires inputs describing background atmosphere temperatures and crosswind velocities as well as aircraft wingspan, and the size of a two-dimensional calculation grid to contain all of the affected zone of the ambient air. The presence of the forest canopy can be expressed as a solid boundary 1 m below the level of mean tree height [Wallace et al (1995)]. Some results for vortex descent and vortex strength decay are shown on Figure 4.3. (Vortex strength is the integrated value of the product of angular rotation rate times the area element in the vortex field, for a single vortex). Vortices decay more quickly for the AG-Truck than for the Grumman TBM, and higher crosswind speed (i.e. higher shear) favors more rapid decay (Fig.4.3a). Vortices sink more rapidly for heavier aircraft with short wings, and a crosswind produces a more rapid descent rate, with apparently an optimum value giving a maximum descent rate (Fig. 4.3b). This last feature has not been reported in practice, to our knowledge. The results agree with the conventional wisdom governing aircraft choice, i.e. a heavy aircraft with short, stubby wings operating at a slow speed is favoured for aerial spraying because these factors favor more rapid wake downwash which means the spray cloud is delivered to the canopy with a greater downward impulse.

The interaction between a crosswind and aircraft wake creates a difference in the behaviour of the upwind wing vortex and the downwind wing vortex. Background vorticity has the same direction (sign) as the aircraft upwind vortex, while the opposite is true for the downwind wing. The additive effect enhances the stability of the upwind vortex and allows it to move closer to the ground, while the downwind vortex decays more quickly and tends to rebound

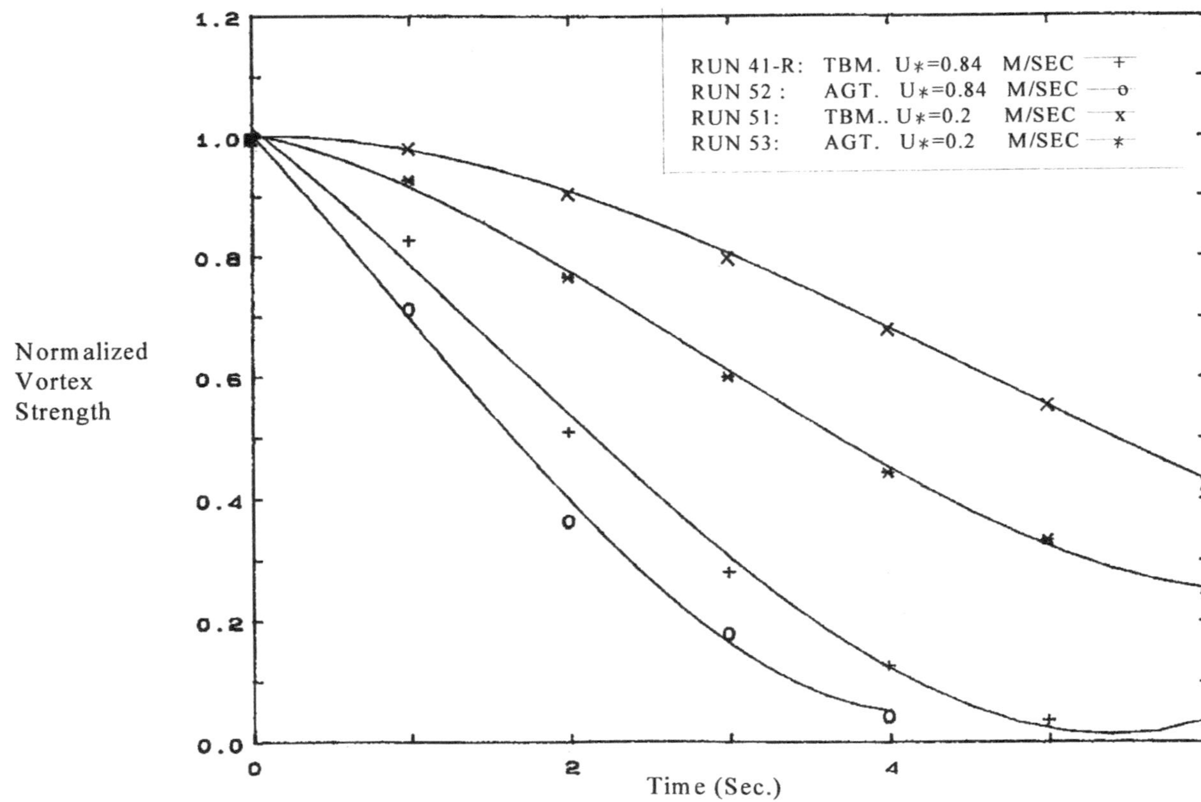

Figure 4.3 a Comparison of Grumman TBM and Cessna 188 (AG-Truck) vortex decay, neutral conditions (TBM at 77 m/s, AG-Truck at 49 m/s airspeed). Copied from Wallace (1988) by kind permission of the author.

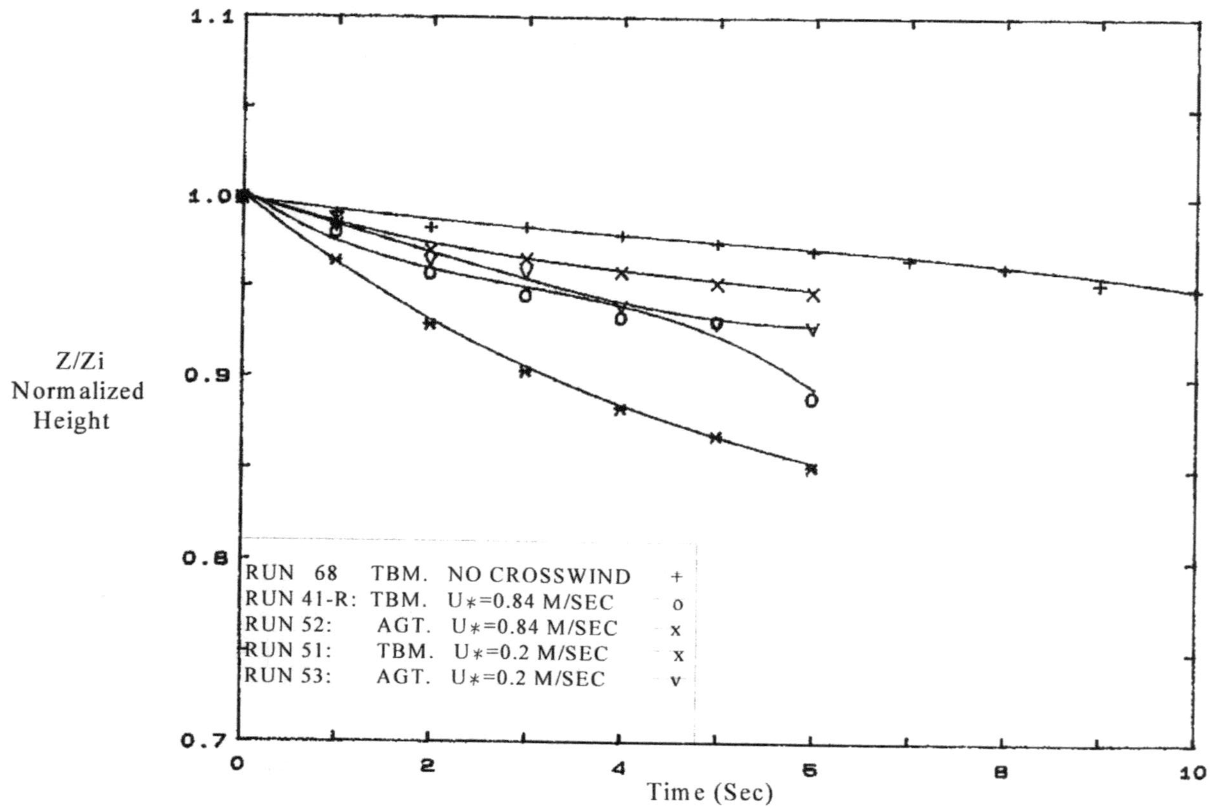

Figure 4.3 b Comparison of Grumman TBM and Cessna AG-Truck (188) vortex descent (TBM at 77 m/s, AGT at 49 m/s airspeed). Copied from Wallace (1988) by kind permission of author.

after it approaches the ground [Bilanin, (1978)]. A field experiment conducted by SERG in New Mexico in 1992 showed that this effect produced less downwind drift beyond 200 m downwind of the flight path for atomizers on the upwind wing than for those on the downwind wing [McCooeye et al (1993)]. This will be further discussed in Chapter 8.

Helicopter wake downwash velocities can be greater than for fixed wing aircraft if the helicopter is operated at low airspeed. Table 4.2 shows a comparison of vortex descent rates for fixed wing aircraft and for a Bell 206 helicopter operating at a range of airspeeds, using equation 4.2.1 [Picot et al (1989)]. At airspeeds below 30 m/s (108 km/hr), the helicopter downwash velocity is greater than that of the TBM and Cessna-188. Field trials were conducted to compare drift from these two aircraft (op.cit.) but the results were confounded by differences in the droplet spectra of the two sprays. That reference reports that for a spray trial over a forested area, there was little difference in the downwind drift beyond 400 m of the spray line, partly due to the fact that the $D_{v.5}$ for the helicopter deposits was 63 micrometers as compared to 83 micrometers for a TBM Teejet spray, for the zone 60 to 150 m downwind of the spray line (finer sprays drift further even when carried by stronger downwash). The atomizer used in the helicopter application was a modified Micronair AU-4000 operating at 14000 rpm (see section 4.5 below).

According to Quantick (1985a), the helicopter operating at speeds above 40 km/hr over large areas offers no advantages over fixed wing aircraft, but below that speed, the high downwash assists the dispersion of large droplets by creating turbulence in the canopy. Lower speed implies longer operating time and higher cost.

TABLE 4.2. Aircraft characteristics and calculated vortex descent rates

A/C Type	Mass kg	Wingspan m	Airspeed m/s	Vortex descent rate m/s
TBM Avenger	8127	16.4	77	0.82
Cessna-188	1904	12.6	49	0.51
Bell 206	1362	10.2	18	1.52
"	"	"	27	1.00
"	"	"	36	0.76
"	"	"	49	0.55

The aircraft WAKE VORTEX model mentioned above will be further discussed in Chapter 6 where it is used in the PKBW spray dispersion and deposition model.

4.3 Droplet Generation and Atomizer Types

Fluid atomizers used on aircraft (Figures 4.4 and 4.5) are usually of the single-fluid type where the fluid containing the active ingredient is forced out of a nozzle or a rotating cone or sieve and is broken up into droplets by the shearing action of the aircraft slipstream. In the case of high speed rotating sieves or wire mesh cages, the rotation increases the relative velocity between fluid and air stream providing for the generation of smaller drops. Droplets of a single, uniform size can be generated in the laboratory under specific closely controlled conditions [Berglund and Liu(1973)]. This is not the case in industrial practice and the measurement of droplet size and frequency is extremely important and a difficult problem (see below). The droplet breakup mechanism varies for these atomizers. [Lefebvre (1989]. For flat fan nozzles, the fluid stream is forced into a flat sheet at the nozzle exit by the internal configuration of the nozzle. Interaction .with surrounding air causes wave-like instabilities to grow, eventually breaking up into ligaments which in turn break up (due to similar instabilities) into droplets. These droplets can eventually shatter into smaller droplets when projected into a high velocity air stream. A similar type of breakup occurs at the edge of spinning discs when fluid is introduced at the centre. The rotating screen devices have internal

Figure 4.4a Tee-jet (on left) and Hollow-cone nozzles, Spraying Systems Inc., Wheaton, Illinois

Figure 4.4b Thru-Valve Boom with radial-jet nozzle in place, one nozzle per 0.30m length of stainless steel streamlined boom section (Waldron Specialties Inc., Ambler, PA)

Figure 4.4c *Microfoil boom and nozzle system (Rhone Poulenc, Lansdale, PA)*

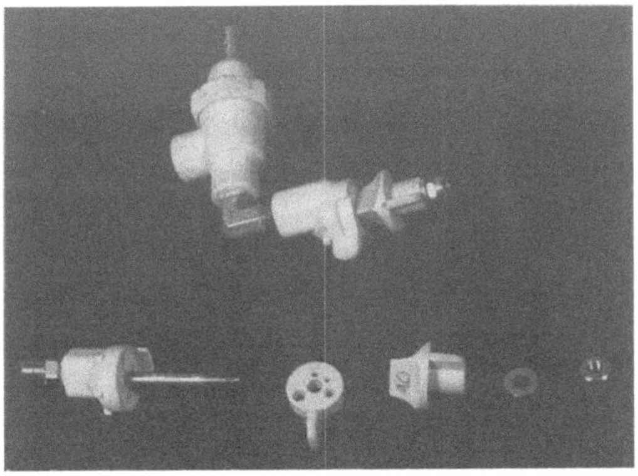

Figure 4.4d *CP nozzle with selectable orifices and deflection plates (CP Products Co., Inc., Mesa, AZ)*

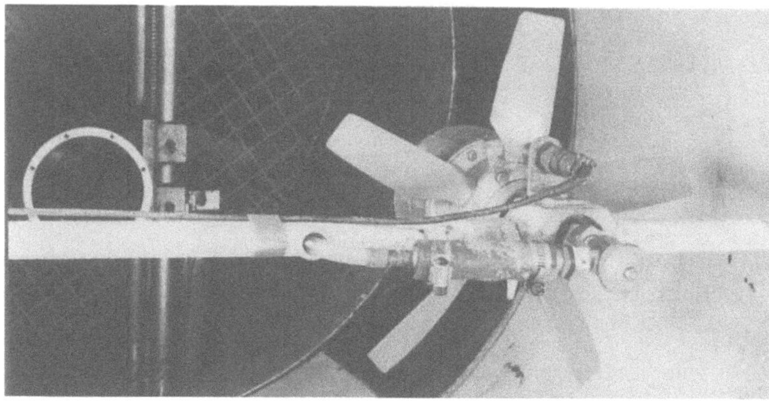

Figure 4.5a Micronair AU4000 rotary atomizer set up for testing in UNB wind tunnel (20mesh sieve 127mm diam. x 79mm long); Micronair Ltd., Isle of Wight, UK

Figure 4.5b Micronair AU5000 atomizer (20mesh sieve 102mm diam. x 79mm long)

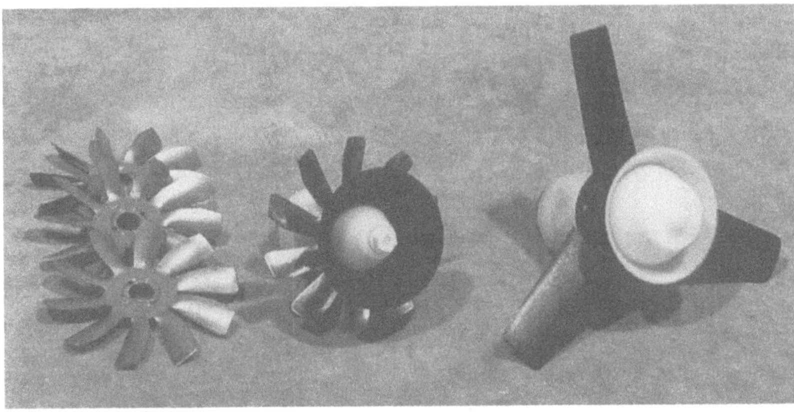

Figure 4.5c Micron X-1 atomizer with different pitch blades (on left, cone ID 48mm, length 54mm, base ID 25mm; Micron Corp., Houston, Texas); Micron X/1 Variable Pitch on right

Figure 4.5d SERG/ONAIR prototype atomizer (custom-built Micronair rotating sieve driven by 3-phase 400 hz motor, US Patent #4,948,050, Can. Patent #2,018,216-4 (Effective Particle Technologies Ltd., Fredericton, N.B.)

fluid distributors which produce a number of radial jets which move outwards to shatter against the rotating screen. In the case of rotating porous sleeves of polyethylene or stainless steel (e.g. the Beecomist atomizers), the fluid emerging from the pores may initially be in the form of ligaments which break up in the usual fashion due to surface instabilities.

4.4 Atomizer Characterization Methods

Accurate information about the size and frequency of droplets produced by an atomizer mounted on an aircraft is essential to the optimal design of a spray application. Experience has shown that atomizers must be tested under full-scale conditions. Current practice is to test the atomizing unit in a wind-tunnel using mounting hardware similar to that used on aircraft. Atomizers are usually mounted under the trailing edge of the aircraft wing, well outside of the wing boundary layer. The velocity of the air stream moving past the atomizer is therefore equal to the aircraft velocity. This is essentially undisturbed ambient air. These two facts define the basic wind tunnel test requirements. Airflow must be turbulence-free and equal to the airspeed of the aircraft. (Note that on some aircraft, atomizers are mounted close enough to the fuselage to to be in the zone where air velocity is higher than the aircraft airspeed [see potential flow around a submerged object, Prandtl and Tietjens (1957)]. This can cause a reduction in droplet sizes at that point of generation). Tunnel length must be sufficient to guarantee that droplet breakup action is finished before drops arrive at the counting and measurement position, and tunnel diameter must be large enough to prevent droplets from striking the tunnel walls ahead of this position. Figure 4.6 shows the wind tunnel used by the authors. The induced draft system is preferred because of the need to have turbulence-free air in a tunnel of manageable size.

Early work on atomizer characterization was based on high-speed flash photography permitting unobtrusive sampling from outside the tunnel. Ingebo and Foster (1957) and Tomney (1978) describe the technique. Subsequent measurements with laser droplet probes have shown that even with the very high speed flash units used (flash persistence of 0.2 microsecs.), severe loss of resolution occurrs for sizes below about 70 micrometers in diameter for experiments done at airspeeds of 40m/s and higher. Photographic methods have been largely replaced by instruments using laser light sources and detectors to measure scattering or occultation of the light by droplets. Recent atomiser characterization work in the authors' laboratory at the University of New Brunswick has been done with the Particle Measuring Systems (PMS) equipment (Figures 4.7 and 4.10) and the Malvern particle spectrometer system (Figure 4.12). The use of these is described separately below.

4.4.1 THE PMS SPECTROMETER

The PMS equipment used by the authors consists of two probes and a data processing unit. The FSSP-100 probe is for a nominal size range (for water) of .5 micrometers to 45 micrometers, and the OAP-260X is for the size range 10 to 620 micrometers. The first is a light-scattering probe, the second a light-imaging probe, and both are single-particle measuring and counting devices. The electronic outputs from these probes are analyzed with a microprocessor-based data analysis unit (PMS model PDS 300).

The use of the FSSP-100 light scattering probe to characterize spray atomizers has been discussed by Picot et al.(1985) and Cerni(1983) and Baumgartner et al (1990) have discussed measurement errors for the probe. This instrument measures the scattered light intensity from a single particle situated in the sample area (the manufacturer supplies the area value, 0.211 mm^2 for the authors' probe) located in the laser beam midway between transmitting and receiving prisms (see Figure 4.7). The light measured is the forward scattered light emitted by the droplet in a hollow cone between 3.3 and 14.3 degrees of the forward direction. The scattering intensity depends on droplet size and refractive index for a non-light-absorbing liquid. For light absorbing liquids, absorptivity must be taken into account. Bontemps (1983) has solved the Mie equations for light scattering from a sphere for the specific probe in question for a number of refractive index and absorptivity values. Figure 4.8 shows some of the results.

The instrument as supplied reports the droplet size spectrum computed on the assumption that the droplets are pure water. Figure 4.9 gives correction factors as a function of refractive index for other liquids (non-light-absorbing).

The data processing unit provides droplet counts for sizes ranging from 0.5 to 45.5 microns (nominally) in fifteen channels with range widths of 3.0 micrometers. Probe sampling time is printed out also along with probe dead time. This is the fraction of sampling time occupied by the computation process, given by an "activity" parameter. Droplet counts passing through the sample area per unit time per unit area are given by the following calculation for channel "i":

$$\frac{(\mathrm{Drops})_i}{\mathrm{mm}^2\mathrm{S}} = \frac{n_i}{S_a \cdot S_t \cdot (1 - A \cdot K)} \tag{4.4.1}$$

Here S_a and S_t are sample area and sample time, A is the activity, N_i is the count registered in channel “i”, and K is an instrument characteristic constant supplied by the manufacturer. This relation is valid when the instrument is in a fixed position. When it is traversed at constant speed along a line perpendicular to the flow direction in a wind tunnel, the calculation must take into account the width

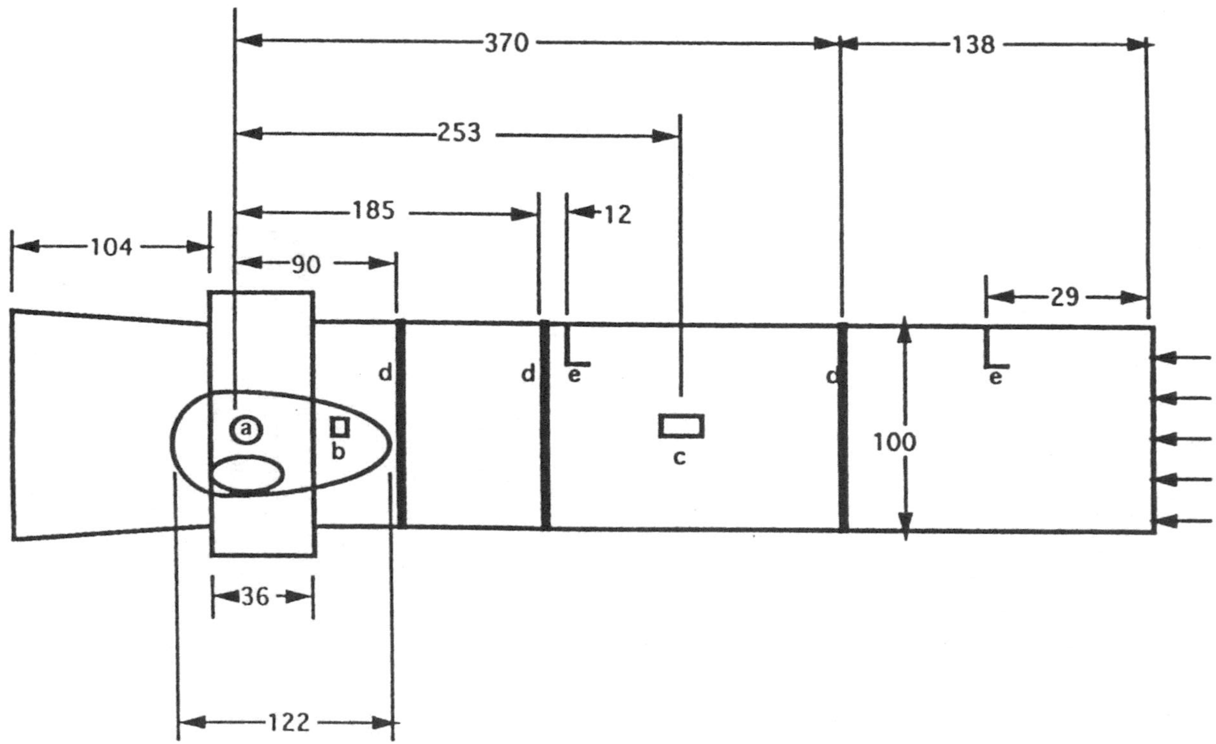

Figure 4.6 Sketch of wind tunnel for atomizer characterization, University of New Brunswick Department of Chemical Engineering. Dimensions in cm, drawing not to scale: a: Malvern measuring position and axial location of vertical traversing system for PMS gear; b,e: Pitot tube measuring positions; c,d: Atomizer mounting positions, equipped with vertical posts.

of the spray plume cross-section sampled by the probe, in this case, a 2-inch(5.08cm) wide strip slightly less than tunnel diameter in length:

$$\frac{(\mathrm{Drop})_i}{s} = \frac{n_i \cdot L \cdot W}{S_a \cdot S_t \cdot (1 - A \cdot K)} \tag{4.4.2}$$

Here W is the spacing between successive lateral traverses across the plume, and L the length of the traverse path. When the traverse sequence encompasses the entire plume cross-section and the traverse spacing W is small enough to ensure representative sampling, the total spray volume flow rate in the complete probe size range can be found by adding the individual channel flow rates:

$$\frac{\mathrm{Vol}}{s} = \sum_i \frac{\pi}{6\, D_i^3} \cdot \frac{n_i \cdot L \cdot W}{S_a \cdot S_t \cdot (1 - A \cdot K)} \tag{4.4.3}$$

When the FSSP-100 size range encompasses the entire droplet spectrum, this calculated flow rate must equal the flow rate supplied to the atomizer, a fact which allows a check on droplet size counts and measurements. Evaporation losses can be checked by calculation and are usually negligible for conditions in the wind tunnel experiments. Calibration procedures for sample area value and droplet size values are required when using this probe, and are reported by van Vliet et al (1989) . When spray droplets are larger than those measured by the FSSP-100, it is replaced by the OAP-260X probe (Figure 4.10) on the wind tunnel probe traversing bracket (see Figure 4.5a). When droplets are larger than 620 micrometers, flash photography is required to complete the droplet spectrum.

The principle of operation of the OAP-260X probe is that droplets passing through a laser beam cast a shadow on an array of photodiodes (Figure 4.10). The shadowed photodiodes are counted electronically and the counts are converted to a size as the droplets pass through the sample area. This probe is a one-dimensional probe and droplets are assumed to be spherical. Extensive testing has shown that the following problems can occur with this probe:

1. Probe electronic reset speed limits the counting frequency to about 35,000 drops per second.
2. Size classes must be calibrated with glass beads of known sizes.
3. Spectrum broadening occurs when the OAP-260X is used to measure a narrow size distribution in the low end of its range.
4. Spatial coincidence of two or more droplets in the sample area at the same time can occur at high droplet fluxes.

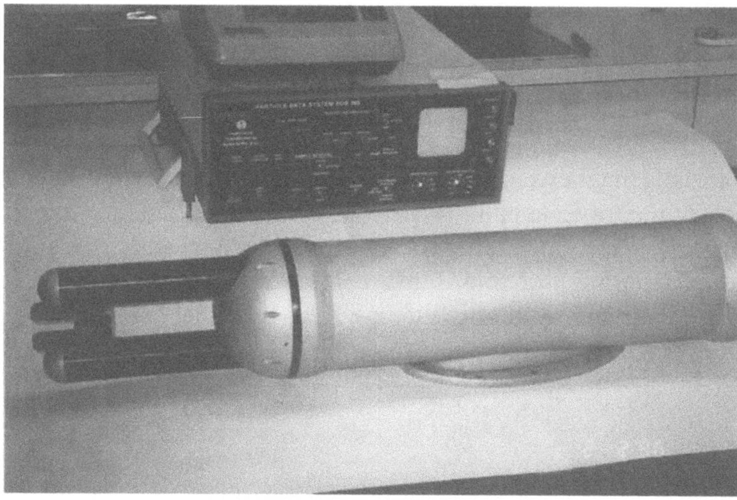

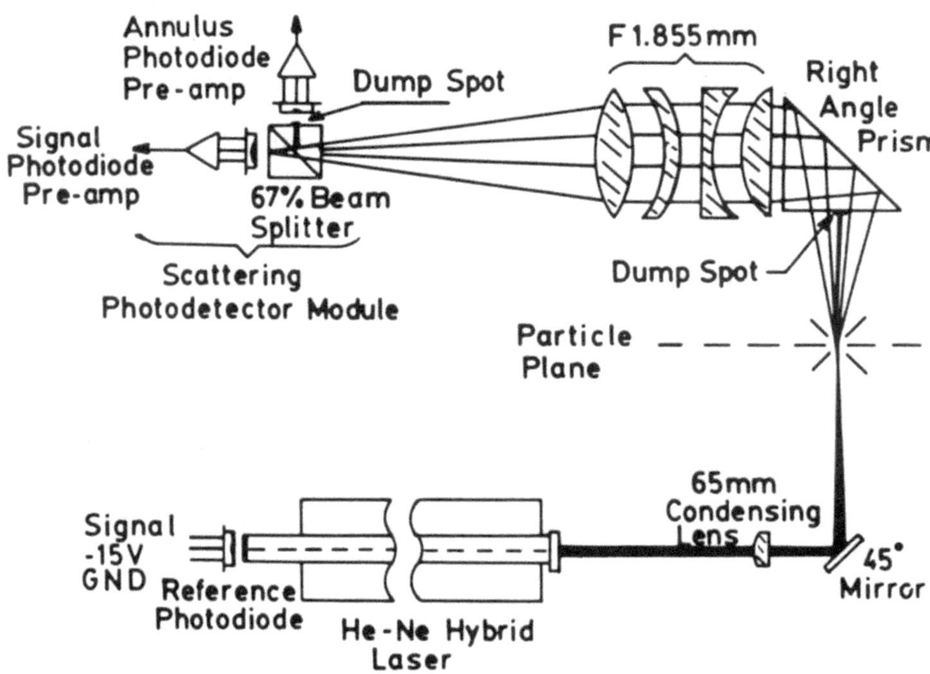

Figure 4.7 Particle Measuring Systems FSSP-100 probe, PDS-300 data processor, and schematic of optical arrangement

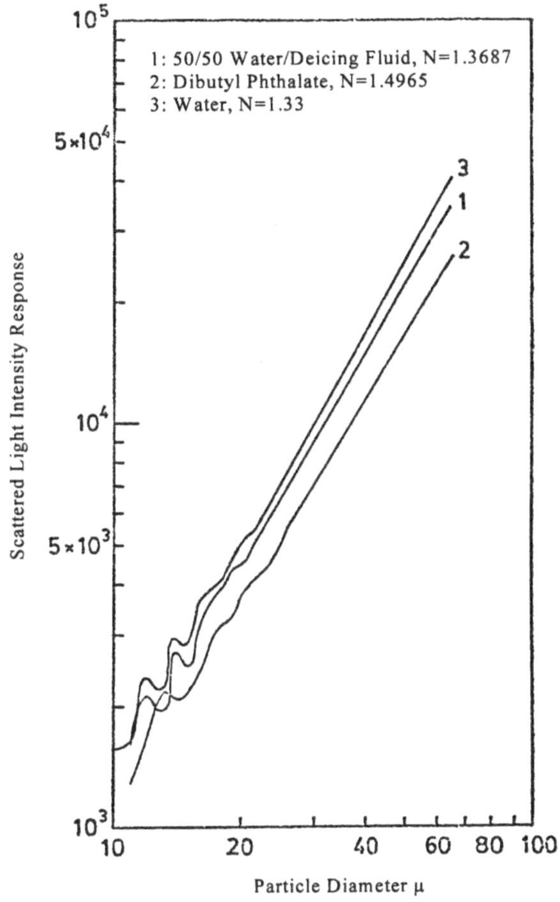

Figure 4.8 Light scattering intensity as a function of droplet size for various refractive index values [Bontemps (1983)].

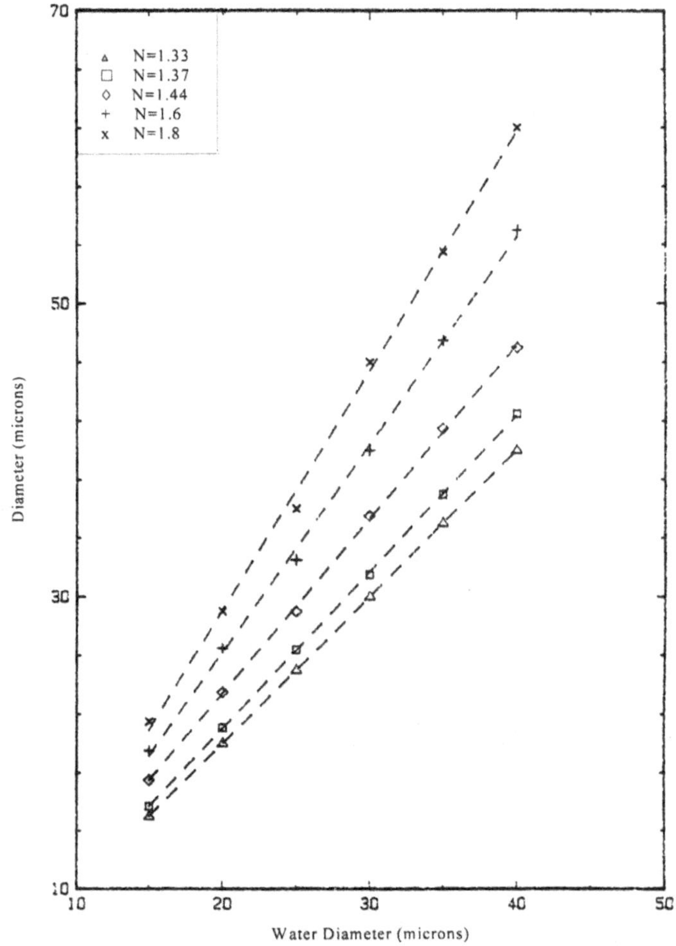

Figure 4.9 Refractive index corrections for FSSP-100 instrument which assumes droplets are pure water [Bontemps (1983)].

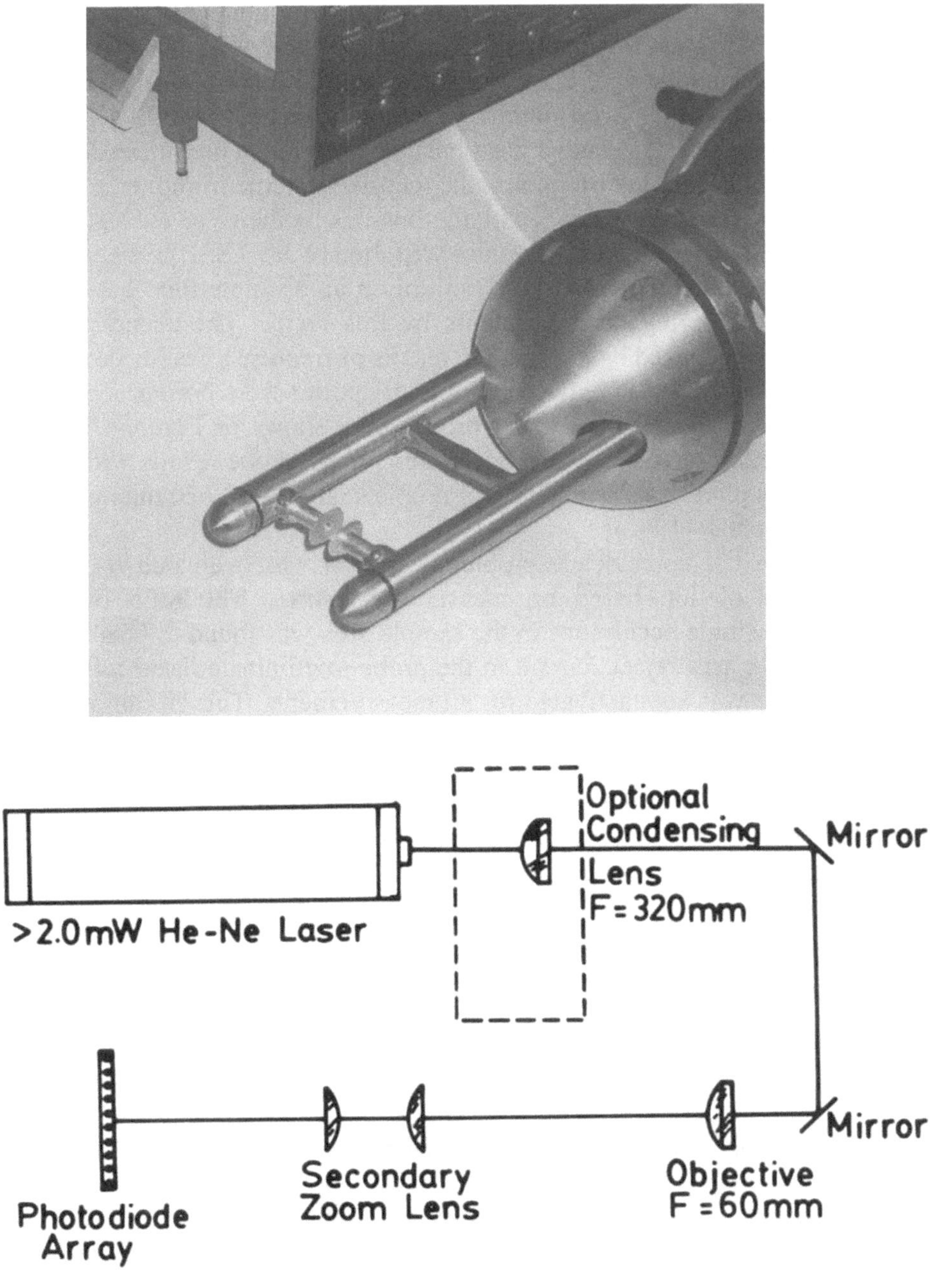

Figure 4.10 Particle Measuring Systems OAP-260X probe and sketch of optical configuration

Item 1. comes into evidence when flow rates are above 5 l/min in a plume cross-section of about $0.1m^2$ when all droplets are smaller than 250 micrometers, for example. The probe will then ignore an unknown number of drops passing through the sample area. Relatively speaking, droplet size distribution will not be affected, but absolute counts on a time basis will be incorrect. Imaging probe results then need to be "spliced" unto the light scattering probe results which are on a corrected time basis. Using a mass balance criterion, a multiplication factor for all OAP counts is found by finding the reported volume from the OAP probe (for sizes above the FSSP range), and dividing this by the value of total volume supplied to the atomizer minus the volume reported by the FSSP probe, all on the same time basis. A corrected OAP spectrum on an absolute time basis is then obtained by multiplying original counts by this factor. The accuracy of the procedure can be checked by plotting the results of frequency versus droplet size and checking the transition between the plotted points from both probes, which should be smooth. The result of such a procedure is shown on Figure 4.11 where a similar procedure was adopted by combining OAP probe results with droplet counts from the photographic method for droplets larger than 620 micrometers in diameter [Picot et al (1990)].

The spatial coincidence problem (item 4) was evaluated by doing a probability calculation based on measured spectra. The ratio of double coincidence to single occurrence in the sample area was found to be about 10^{-6}. There is a "bulk area reject" circuit in the probe to eliminate these coincidence events, and this was kept activated for all measurements. This circuit calculates the transit time of droplets through the sample volume, converts this to a droplet diameter, and compares this with the value obtained from shadowed photodiode counting. When the second value is significantly different from the first, the droplet count is rejected. Because of the perceived low probability of coincidence events, a detailed analysis of the accuracy of this circuit was not performed. It is to be noted that the initial OAP volume calculation differs from equation 4.4.3 as follows:

$$\frac{Vol}{s} = \Sigma_i \frac{\pi}{6\ D_i^3} \cdot \frac{n_i \cdot L \cdot W}{(S_{a,i} \cdot S_t)} \qquad (4.4.4)$$

For the OAP-260X, sample areas (represented by $S_{a,i}$) depend on the size channel, of which there are 64. These are supplied by the manufacturer.

Hovenac et al (1985) and van Vliet et al (1989) have done detailed studies of PMS imaging probe accuracy and calibration, with respect to sample areas and sizes. Areas are supplied by the manufacturer and Figure 4.12 shows the size calibration curve for the instrument used by the authors.

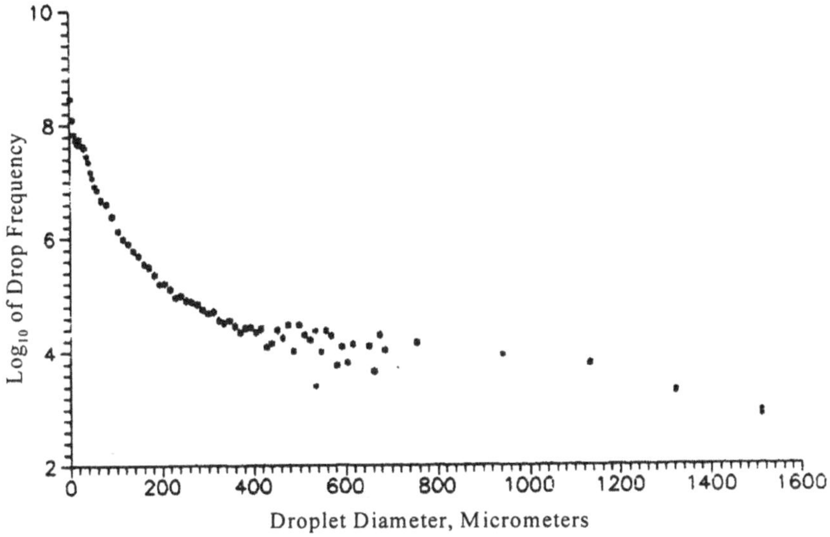

Figure 4.11 Droplet frequency (drops per μm of diam. per sec.) For a 2.6 l/min flow of a liquid with viscosity 1.8 mNs/m^2 and surface tension 48.4 mN/m, for a rearward -facing radial array of 12-1.8 mm ID tubes (a TVB nozzle, Fig. 4.4b). Photographic data corrected to temporal basis. Copied from Picot et al (1990) by the kind permission of ASTM. (FSSP data 0.5 to 55 μm, OAP data 55 to 620 μm, photographic data >620 μm).

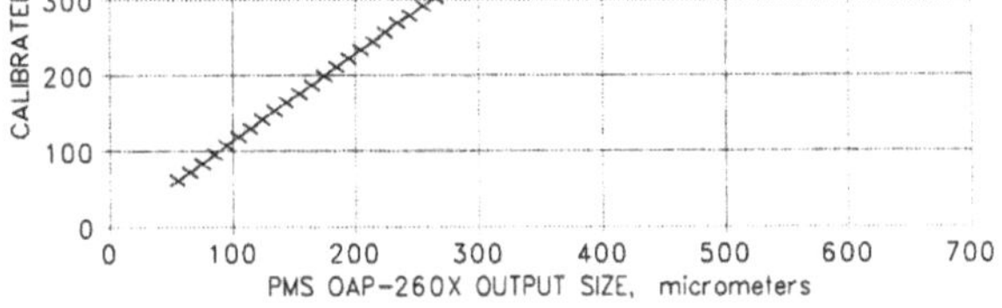

Figure 4.12 Calibrated size vs instrument output size, OAP-260X probe

4.4.2 THE MALVERN SPECTROMETER

This instrument, manufactured by Malvern Instruments, Malvern, U.K., is a "droplet cloud" probe. It functions by measuring the Fraunhofer diffraction pattern formed by a number of particles illuminated by a laser beam (Figure 4.13). Brown and Felton (1985) summarize the principles used to obtain particle size distribution and concentration with this device. The light energy variation with radius on the focal plane of the diffraction pattern is used to measure size distribution, and the degree of obscuration of the light beam by the particle cloud is used to obtain particle concentration. The light energy contained within any ring on the focal plane bounded by radii s_1 and s_2 due to a particle of radius a_i is given by:

$$L_{s1,s2} = C''\sum\frac{W_i}{a_i}\left\{J_o^2\left(\frac{ka_is_1}{\Lambda}\right) + J_1^2\left(\frac{ka_is_1}{\Lambda}\right) - J_o^2\left(\frac{ka_is_2}{\Lambda}\right) - J_1^2\left(\frac{ka_iS_2}{\Lambda}\right)\right\} \quad (4.4.5)$$

Here W_i is the combined mass of the drops in size class i of radius a_i, k is the laser light wave number (equal to $2\pi/\lambda$), Λ is lens focal length, and J_o and J_1 are the zero and first order Bessel functions of the first kind. C'' is a constant which can be evaluated. The summation is carried out over M size ranges within which size is assumed uniform. With measured values of $L_{s1,s2}$ available, it is possible to solve for the relative mass (and volume) distribution for these M ranges, with M having a maximum value of 32 in the case of the Malvern instrument. This is accomplished by the Malvern software using the "model-independent " software option. These data can be converted to number fraction, and the measurement is based on a spatial sample. Measurements are made at a number of radii on the focal plane according to the expected size distribution, where the first maximum radius s_i of the energy distribution is found at $2\pi \cdot a_i \cdot s_i/(\lambda\Lambda) = 1.357$ for a size range with a mean radius a_i. The Fourier transform lens focusses the undiffracted light onto the centre spot of the photocell system and the instrument reads the ratio of light intensity with and without the drops present. This is called the obscuration, Ob. When the size distribution has been obtained from above, i.e. number fraction f_i in terms of a_i for a spatial volume, the calculation of volume flux requires a knowledge of total droplet numbers and their velocity. When droplet density is low (the case for most situations in agriculture and forestry spraying), droplets in the laser beam are not shadowed by other droplets, and there is an insignificant level of scattering from one droplet to another [Goguen (1994), Palopski and Kankunnen (1993)]. All droplets are then illuminated by the same light intensity, and the obscuration of emitted light can be written as follows:

$$Ob = \left(\frac{N_{tot}}{A_{beam}}\right) \Sigma\, f_i\, C_{sca,i} \quad (4.4.6)$$

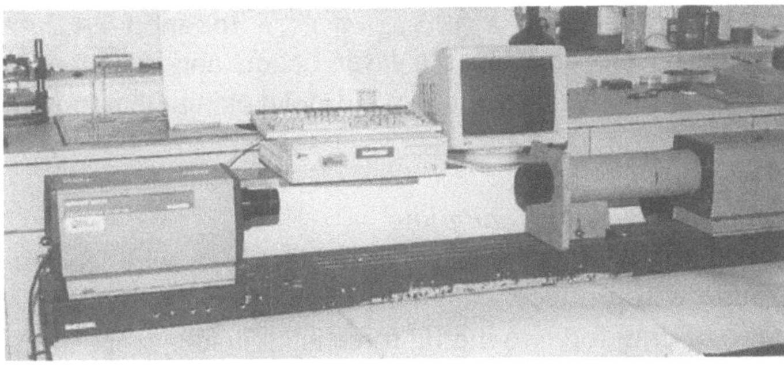

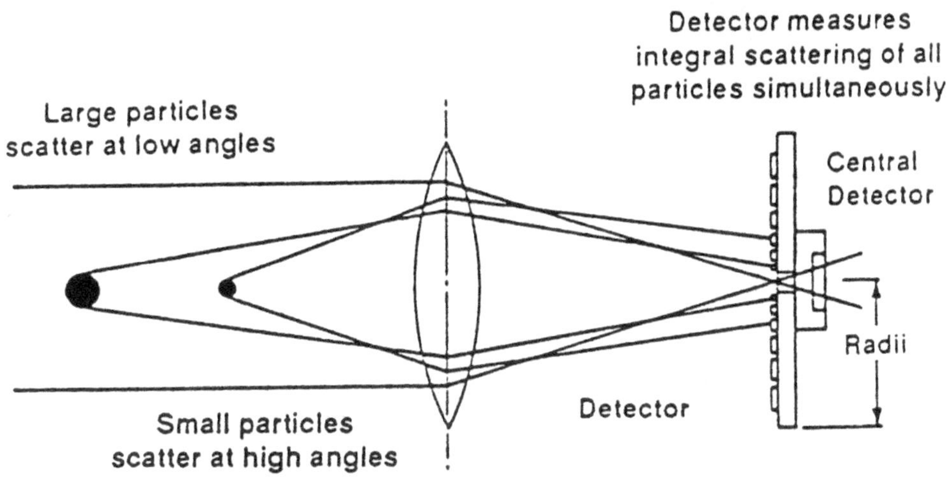

Figure 4.13 Photograph of Malvern spectrometer system, and sketch of optical configuration. The optical bench permits rearrangement of optical components to provide measurements of particles from 1.2 to 1880 micrometers in six size ranges. In the UNB wind tunnel, pedestals on either side (and outside) of the test section replace the optical bench.

Here Ob is the obscuration, available as output from the Malvern instrument, N_{tot} the total number of droplets in the laser beam, and $C_{sca,i}$ is the scattering cross-section for each droplet size, with the summation carried out over all size classes M present. A_{beam} is the laser beam cross-section. For droplets larger than 4.5 micrometers, the value of $C_{sca,i}$ is equivalent to $2\pi a_i^2$ [van de Hulst (1981)]. The optical bench permits rearrangement of optical components to provide measurements of drops ranging from 1.2 to 1880 micrometers in diameter.

When droplet velocities U_i are uniform (and known) for each size class, droplet number flux through the laser beam is as follows, for size class i:

$$N_i = \frac{N_{tot}\ f_i\ U_{d,i}\ D}{A_{beam}} \tag{4.4.7}$$

Here D is the laser beam diameter $(4\ A_{beam}/\pi\ )^{0.5}$. Equation 4.4.7 gives the flux of droplets for each size class in the laser beam, and volume flux is easily computed by multiplying this by volume per droplet using the nominal droplet size for each size class.

Spatial Segregation and Plume Sampling

In general, rotary atomizers and nozzle atomizers with jets entering the air stream at an angle will produce a spatially-segregated droplet size distribution. Because the ratio of drag force to inertia force is inversely proportional to droplet size, larger drops are flung out further radially from a rotating atomizer than smaller drops, and larger drops from an angled jet will project further away across the air stream than smaller drops (see below). To get overall spray plume characteristics, a traverse is required, i.e. the laser beam must be relocated step-wise across the spray plume and measurements done at each location. Traverse step size must be chosen to ensure adequate sampling of the complete plume. In this case, the N_i of equation 4.4.7 becomes $N_{i,j}$ denoting droplet number flux for size class i at position j, and the total volumetric flow rate for the complete plume is given by:

$$F_{tot} = \Sigma_i\ \Sigma_j\ N_{i,j}\pi\left(\frac{d_i^3}{6}\right)\frac{\Delta Z}{D} \tag{4.4.8}$$

Summations are on all j and i for all positions and size classes (d_i is twice the radius a_i) present in the spray. The equation is valid for ΔZ values equal to or greater than D.

Equation 4.4.8 allows an estimate of the accuracy of droplet size and frequency measurement, simply by comparing the result with the volumetric flow rate supplied to the atomizer during the test. **It is important to note that the concentration of drops calculation done in the Malvern output software assumes a Beer's law light absorption dependency in the laser beam. This can only be correct when droplet size distribution is uniform along the light beam. This is not the case in most agricultural spray plumes, even in wind tunnels (see below). Therefore, the above method is the only one available for volume flux computation, and it is only successful when the assumption of dilute spray is valid. It will be shown below that very good volumetric flow balances result for situations when individual traverse obscuration are less than 20%.** In cases where obscurations measured by the Malvern instrument are 50% and higher, Palopski and Kankkunen (1993) present correction methods when log-normal and Rosin-Rammler droplet size distributions can be assumed, but at this time, no correction method has been developed for the model-independent method discussed above.

4.4.3 OTHER SPECTROMETER DEVICES

The phase Doppler particle size measurement device is a single-particle device, and uses the principle that the relative phase shift of the light scattered by reflection at different angles is proportional to the droplet size [Bachalo(1988)]. Such equipment is manufactured by Aerometrics, Inc., Mountain View, CA 94042, and by Dantec Measurement Technology Inc., Mahwah, NJ 07430. These instruments provide droplet velocity measurements simultaneously as droplets are measured. A comparison of size distributions obtained by the PMS, Malvern, and Aerometrics devices shows good agreement between the three [Young and Bachalo(1988)]. The phase Doppler type of equipment appears to be more cumbersome to use in a wind tunnel when full plume cross-section surveys are required.

Flash photography has largely been replaced by the opto-electronic methods mentioned above, but has some use in getting information on the approximate maximum droplet size in a spray when capital for purchasing equipment is limited.

4.5 Atomizer Droplet Spectrum Data

The USDA Forest Service has compiled drop size data obtained in wind tunnel tests [Skyler and Barry (1991)] done at the University of California at Davis, in the laboratories of Professors N. B. Akesson and W.E Yates. They utilized a PMS OAP-2D-GA-1 spectrometer, a two-dimensional imaging device covering the size limits 28 to 2062 µm in 64 channels [Yates et al (1985)]. The data include over 200 combinations of atomizer, fluid, flow rate, and airspeed. Picot et al (1989) and Goguen (1994) give data on similar atomizers, providing values of $D_{v.1}$, $D_{v.5}$, and $D_{v.9}$ obtained with the PMS equipment and methods described in section 4.4. A comparison of tests done on D8-46 hollow-cone nozzles is shown on Figure 4.14. The "Roundup" simulant from the Goguen (1994) data was a 9% solution of isopropanol in water with a surface tension of 43.5 dyne/cm and a viscosity of 1.8 cp at 20°C. The USDA Forest Service data gives the following values for $D_{v.1}$, $D_{v.5}$, and $D_{v.9}$: 141, 261, and 443 µm. The Goguen data gives 15, 143, and 271 µm for these values on the lower surface tension fluid. The expected differences caused by the different fluid properties are minor compared to the differences in the reported values. Goguen (1994) carried out an extensive investigation on possible measurement errors using both the PMS equipment and Malvern equipment with the D8-46 nozzle as test subject. Figure 4.15 summarizes the results from a multiple-position Malvern measurement using the method described in Section 4.4. The values of $D_{v.1}$, $D_{v.5}$, and $D_{v.9}$ from Figure

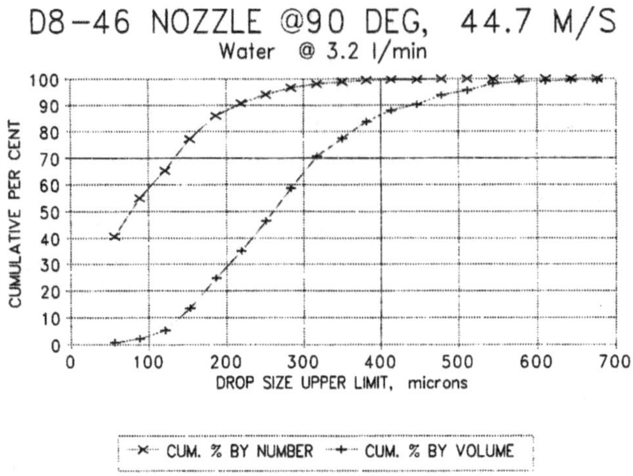

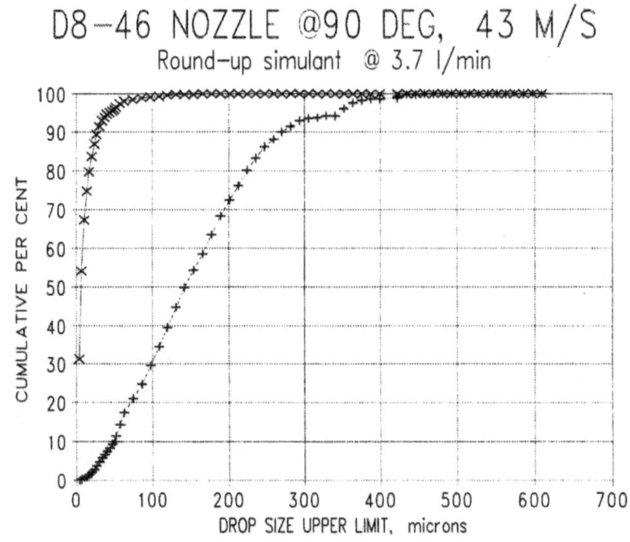

Figure 4.14 Results for similar tests reported by Skyler and Barry (1991) - top; and Goguen (1994) - bottom (PMS system).

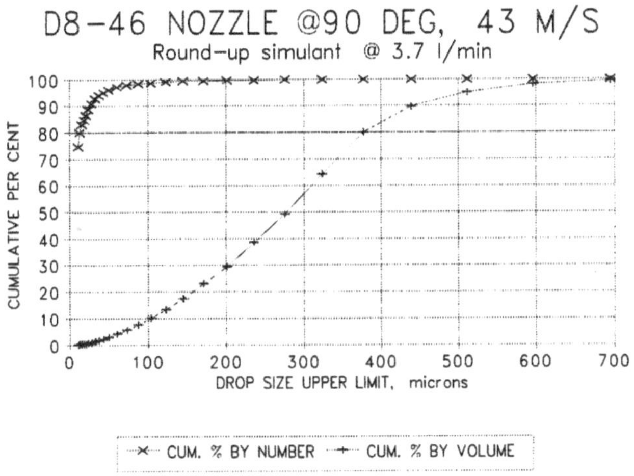

Figure 4.15 Goguen (1994) characterization results using Malvern multiple- position traverse and calculations as described in section 4.4

4.15 are 29, 278, and 440 μm respectively. The spray plume total measured volume was 91% of the volume supplied to the atomizer. (Improvements in flow rate control reduced this mass balance closure error in subsequent tests from the above 9% value to less than 5%). The UNB PMS number frequency data and the Malvern number frequency data are replotted on Figure 4.16 [Goguen et al (1996)]. This plot shows good agreement between the two sets of data for drop sizes of about 230 μm and lower, although the PMS values are systematically slightly higher. Frequency values from the PMS system are much lower at higher diameter values. The curve-matching routine for FSSP and OAP probes (section 4.4) comes into effect at a diameter of 50 μm in this case and does produce an upward bump in the curve, which is indicative of distortion in the OAP data. Because of the sharp drop-off in the OAP counts at sizes greater than 230 μm, the cumulative volume fraction values have suffered a large error. Since the PMS FSSP probe was calibrated for size, sample area, and probe dead time, the error is undoubtedly in the performance of the OAP probe. It is possible that at high count rates, the passage of the probe through regions of rapidly changing drop size distribution (as in this case with a nozzle at 90 degrees to the airflow - see below) creates a distortion in the droplet spectrum because probe dead time does not omit a representative fraction of the whole spray plume. It is therefore concluded that the data of Picot et al (1989) on hollow cone nozzles are in error (see Table 3, p.756 of that paper). As well, the USDA Forest Service data of Figure 4.14 seem to report an erroneously small number of small droplets. The document does not mention the mass balance criterion to verify measurement. A paper describing the functioning of the OAP device [Knollenberg (1970)] is generic and does not deal

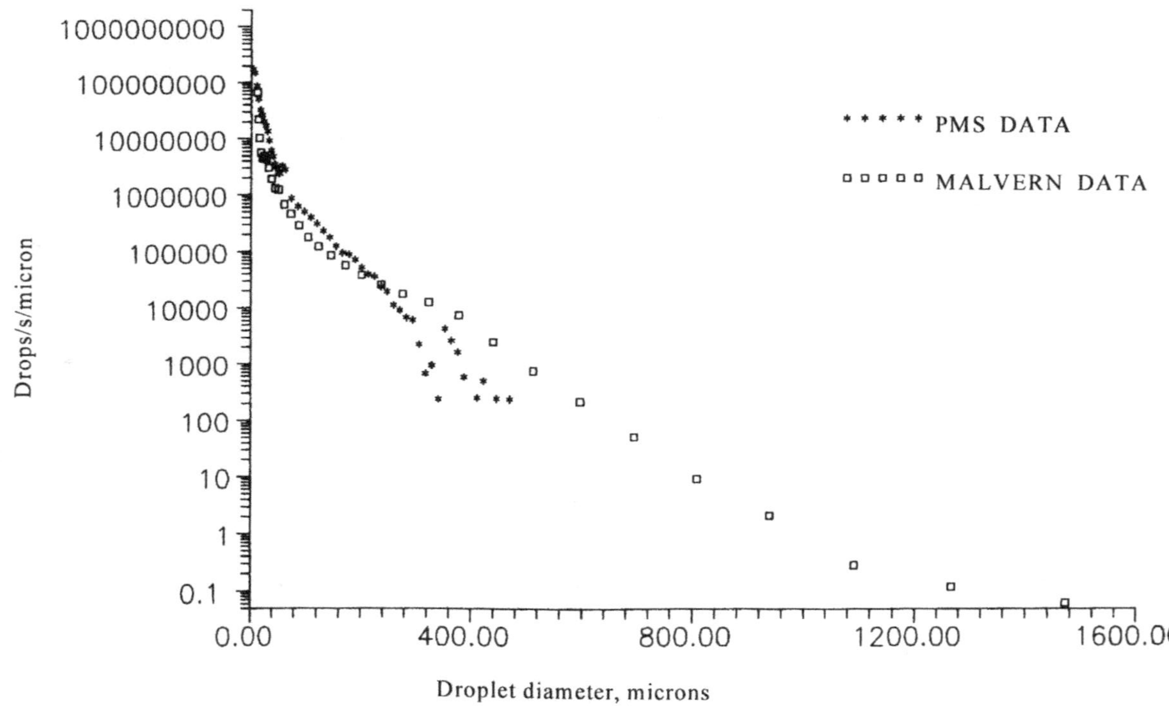

Figure 4.16 Droplet number frequency vs diameter, comparison between PMS and Malvern multiple-position traverses, D8-46 nozzle at 90 degrees to the air flow at 43 m/s, with 3.7 l/min of "Round-up" simulant

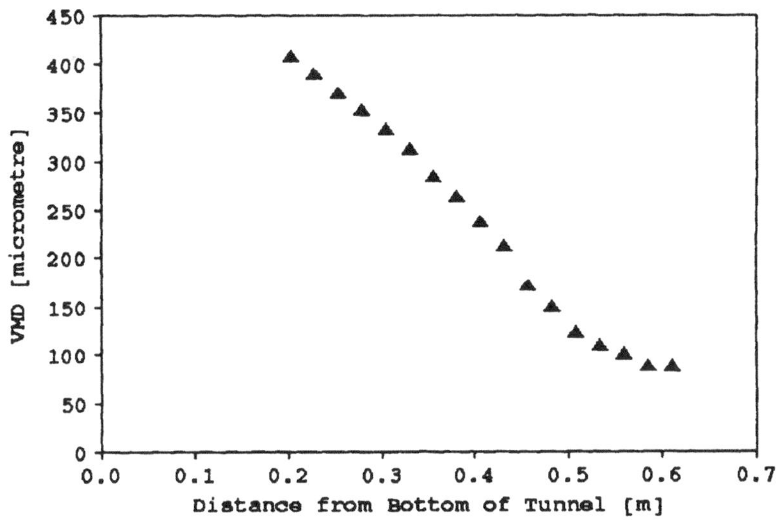

Figure 4.17 Volume median diameter (VMD or $D_{v.5}$) as function of vertical position of the D8-46 nozzle at 90 degree orientation to air flow. The Malvern laser beam traverses the wind tunnel horizontally at a distance of 0.55 m from the bottom (floor) of the wind tunnel and 3.7 m downwind of the nozzle.

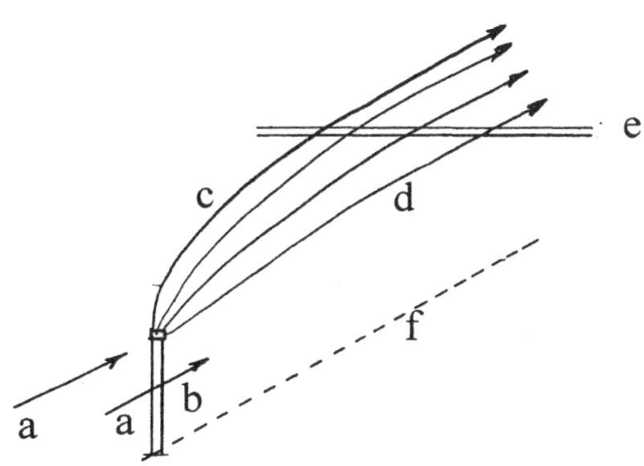

Figure 4.18 Relative positions of wind tunnel components as they relate to Figure 4.17: a -air flow vectors; b -nozzle mounting post with vertical adjustment; c -pathline of largest drops; d -pathline of smallest drops; e -Malvern laser beam at fixed horizontal position and 3.7 m down- wind of the nozzle; f - downwind orientation of tunnel floor

with the 2-D probe specifically. The device appears to be subject to the same probe "dead time" problems as the 1-D probe.

Goguen (1994) has demonstrated the droplet size segregation problem when nozzles such as the D8-46 are oriented at right angles to the atomizing air flow direction. This is shown on Figure 4.17 for the test conditions discussed on Figures 4.15 and 4.16. Measurements directly downwind of the nozzle show a $D_{v.5}$ of 100 μm or slightly less, while measurements taken 0.35 m above the nozzle (i.e. with nozzle 0.20 m above tunnel floor) show a $D_{v.5}$ of 400 μm or slightly higher. Figure 4.18 shows the relevant positions of nozzle, laser beam, tunnel floor, and air flow direction. The calculation methods for a multiple-position Malvern spectrometer traverse described in section 4.4 are necessary to convert the 17 separate measurements to a spectrum representing the entire plume, and also to a droplet volume flux for purposes of checking overall measurement accuracy. The vertical step sampling interval was 1.25 cm in this case. None of the individual sampling positions can be used alone to give representative data of the total spray plume. However, in the case of the SERG/ONAIR high speed rotating atomizer, it was found that a single measurement through the plume center line gave values similar to those given by the UNB PMS system on the complete plume (Figure 4.19), probably because of the very small droplets generated and the vigorous mixing occurring in the immediate wake of the atomizer. **It is erroneous to conclude from this that Malvern spectrometer characterizations can be done with a single "shot" through a plume center-line!**

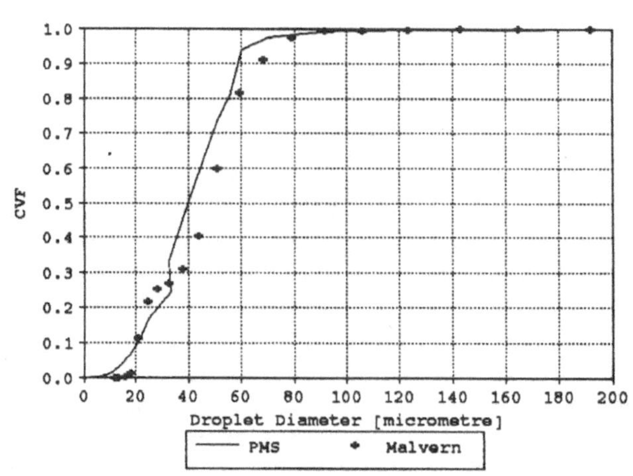

Figure 4.19 Cumulative volume fraction vs drop size for the SERG/ONAIR atomizer rotating at 14,700 rpm with fluid at 2.0 l/min (a 9% isopropanol in water solution); air velocity 40 m/s.

The work by Goguen (1994) includes details of a size calibration procedure for the Malvern spectrometer.

A series of flux-based atomizer characterizations done with the Malvern device on insecticide and herbicide nozzles is presented in Appendix II [Picot and Goguen (1995)].

4.6 Correlations for Atomizer Emission Characteristics

The wide range of conditions, designs, and fluid types used for forestry atomization has made it impossible to develop useful general correlations for predicting droplet spectrum mean diameters. Published correlations for apparently similar situations differ widely. For example, correlations for the Sauter mean diameter (SMD, equivalent to $\Sigma N_i D_i^3 / \Sigma N_i D_i^2$, where N_i refers to the number of drops of size D_i and summation is over all drops) are given for liquid jets in a crossflow of air by Ingebo and Foster (1957) and by Kihm et al (1995). In the first of these, the SMD is proportional to $V_R^{-0.75}$, and in the second, it is proportional to $V_g^{1.1}$, where V_R is relative velocity between fluid and air, and V_g is air velocity. For coaxial flow of liquid and air, Mansour and Chigier (1995) report a dependency of SMD proportional to V_g . For a variety of nozzles and rotating atomizers, Picot et al (1989) report that the $D_{v.5}$ is proportional to $V_R^{-1.6}$, in a correlation which varies with the data by up to 35%. Akesson and Gibbs (1990) report different power dependencies on wind tunnel air velocity, namely -0.40, -0.58, and -1.24 for fan, hollow cone, and jet nozzles. Their approach of developing correlations for each atomizer type seems to be a good practical solution. Additional wind tunnel tests for specific cases are required. Accurate results for the complete droplet size range are necessary in view of the need to predict droplet behaviour downwind of an aircraft application. This situation has been addressed for some years now by the Spray Drift Task Force, a committee of representatives of the agricultural chemical industry in the United States. At time of writing, no reports have been published on their atomizer characterization results.

4.7 Dry Material Dispensing Devices

Aerial application of solid formulations has had limited use in forestry. Cochran and Boring (1988) describe the application of a granular herbicide to control herbaceous weeds, brush and hardwood plants in a conifer plantation. The formulation ("Pronone", manufactured by Pro-Serve Inc., Memphis, TN and DuPont) consists of the chemical hexazinone coated onto the irregular surface of calcined montmorillonite clay granules. The authors describe application of this material by helicopter using a modified Simplex Seeder. This device consisted of a 15-foot boom on each side of the helicopter into which the granular material was metered into an air stream blown through the boom. Each boom had five outlet ports oriented to the rear. The metering devices consisted of variable-speed rotary valves.

Barry et al (1974) describe a field test of a carrier dust (hygroscopic silicate particles coated with a liquid chemical pesticide) applied with a Cessna Agwagon equipped with a Transland Swathmaster spreader. The $D_{v.5}$ of the formulation was 13 μm, and the insecticide (Zectran FS-15 from Ciba-Geigy) application rate was 0.36 lb per acre (0.40 kg/ha). Mortality to the target insect (western spruce budworm) was very low apparently due to the low level of deposit on conifer foliage.

Quantick (1985a and 1985b) gives a review of solids application by aircraft with details of the application equipment. Figure 4.20 is a generic sketch of a solids spreader which uses the aircraft slipstream (e.g. the Transland

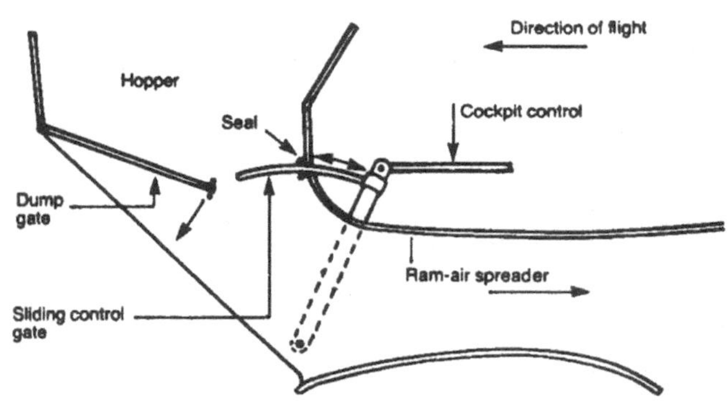

Figure 4.20 Typical slip-stream solid particle distributor. The downstream end is usually fan-shaped to assist in the lateral spreading of the formulation. (Copied from Quantick (1985b) by kind permission of Blackwell Science Ltd, publishers, Oxford, Eng.)

Swathmaster). Figure 4.21 shows a version of a similar device designed at the Cranfield Institute of Technology, in Cranfield, England.

The material presented by Quantick in the above references gives a comprehensive account of the flow of granules through metering devices as well as information on the swath widths achievable with a variety of dispensing devices. The reader is directed to those publications for the details.

Figure 4.21 Cranfield designed S.B.-1 spreader fitted to a Piper Pawnee aircraft. (Copied from Quantick (1985a) by kind permission of Blackwell Science Ltd., publishers, Oxford, England)

References

Akesson, N.B., and R.E. Gibbs (1990): Pesticide drop size as a function of spray atomizers and liquid formulations, Pesticide Formulations and Application Systems, **10**, ASTM STP 1078, L.E. Bode, J.L. Hazen, and D.G. Chasin, eds. ASTM, Philadelphia, PA., pp170-183.

Bachalo, W.D. (1988): The phase Doppler method: Analysis and application, Optical Particle Sizing, Theory and Practice, G. Gouesbet and G. Grehan eds., Plenum Press, New York, pp 283-299.

Barry, J.W., M. Tysowsky, R.E. Ekblad, and W.M. Ciesla (1974): Carrier dusts- feasibility for forest insect control, Trans. ASAE, **17**, pp 645-650.

Baumgardner, D., W.A. Cooper, and J.E. Dye (1990): Optical and electronic limitations of the forward-scattering probe, Liquid Particle Size Measurement Techniques: **2**, ASTM STP 1083, E.D. Hirleman, W.D. Bachalo, and P.G. Felton eds., ASTM, Philadelphia, PA, pp 115-127.

Berglund, R.N., and B.Y.H. Liu (1973): Generation of monodisperse aerosol standards, Env. Sci. Tech. **7**, pp 147-153.

Bilanin, A.J., M.E. Teske, and G.G. Williamson (1977): Vortex interactions and decay in aircraft wakes, A.I.A.A. Journal, **15**, no.2, pp 250-260.

Bilanin, A.J., M.E. Teske, and J.E. Hirsh (1978): Neutral atmosphere effects on the dissipation of aircraft wakes, A.I.A.A. Journal, **16**, no.9, pp 956-961.

Bontemps, X. P-M. B. (1983): The Optimization of Aerial Spray Atomizers, M.Sc. thesis, Dept. of Chemical Engineering, University of New Brunswick, Fredericton, N.B.

Brashears, M.R., and J.N. Hallock (1974): Aircraft wake vortex transport model, J. Aircraft, **11**, no.5, pp 265-272.

Brown, D.J., and P.J. Felton (1985): Direct measurement of concentration and size for particles of different shapes using laser light diffraction, Chem. Eng. Res. Des., **63**, pp 125-132.

Cerni, T.A. (1983): Determination of size and concentration of cloud drops with an FSSP, J. Clim. and Appl. Met. **22**, pp 1346-1355.

Cochran, R.J., and L.E. Boring (1988): Aerial application of a granular herbicide, Pesticide Formulations and Application Systems, **8**, ASTM STP 980, D.A. Hovde and G.B. Beestman eds., ASTM, Philadelphia, PA, pp 283-287.

Donaldson, C. duP. (1973): Construction of a Dynamic Model of the Production of Atmospheric Turbulence and the Disperal of Atmospheric Pollutants, American Meteorological Society Workshop on Meteorology, D.A. Haugen (ed.), Science Press, pp 313-390.

Elliott, J.G., and B.J. Wilson (1983): The Influence of Weather on the Efficiency and Safety of Pesticide Application, British Crop Protection Council Publications, Occasional Publication No. 3, London, England.

Goguen, R.J. (1994): Atomizer Characterizations with the Malvern 2605LC Particle Size Analyzer, M.Sc. Thesis, Dept. of Chemical Engineering, University of New Brunswick, Fredericton, N.B.

Goguen, R.J., B.A.J. Saunders, and J.J.C. Picot (1996): Flux-based measurements with the Malvern spectrometer; paper submitted for the 1996 ASAE annual meeting, Phoenix, Arizona; Dept. of Chemical Engineering, University of New Brunswick, Fredericton, N.B.

Hallock, J.N., and W.R. Eberle (eds., 1977): Aircraft Wake Vortices: A State of the Art Review of the United States R & D Program. Report No. FAA-RD-77-23, Feb. 1977.

Hovenac, E.A., E.D. Hirleman, and R.F. Ide (1985): Calibration and sample volume characterization of PMS optical array probes, International Conference on Liquid Atomization and Spray Systems, **2**, July 1985.

Ingebo, R.D., and Foster, H.H. (1957): Drop-Size Distribution for Cross-Current Break-up of Liquid Jets in Air Streams, NACA TN 4087.

Kihm, K.D., G.M. Lyn, and S.Y. Son (1995): Atomization of cross-injecting sprays into convective air stream, Atom'n and Sprays, **5**, pp417-433.

Knollenberg, R.G. (1970): The optical array: An alternative to scattering or extinction for airborne particle size distribution, J. Appl. Met., **9**, pp 86-103.

Lefebvre, A.H. (1989): Atomization and Sprays, Hemisphere Publishing Corporation, New York.

Mansour, A., and N. Chigier (1995): Air-blast atomization of non-Newtonian liquids, J. Non-Newt. Fl. Mech., **58**, pp 161-194.

McCooeye, M.A., R.S. Crabbe, R.E. Mickle, A. Robinson, E.B. Stimson, J.A. Arnold, and D.G. Alward : Strategy for Reducing Drift of Aerially Applied Pesticides, report from National Research Council Canada, Institute for Environmental Chemistry, Ottawa, ca 1994.

Palopski, T, and A. Kankkunen (1993): Multiple scattering and size distribution effects on the performance of a laser diffraction particle sizer, Atomization and Sprays, **3**, pp 311-320.

Picot, J.J.C., X. Bontemps, and D.D. Kristmanson (1985): Measuring spray atomizer droplet spectrum down to 0.5 μm size, Trans. A.S.A.E. **28**, pp 1367-1370.

Picot, J.J.C., and R.J. Goguen (1995): Spray Atomizer Characterizations- Final report, Dept. of Chemical Engineering, University of New Brunswick, Fredericton, N.B.

Picot, J.J.C., D.D. Kristmanson, M.S. van Vliet, and C. Cormier (1989): The Use of Controlled Aircraft Vortex and UULV Methods to Improve the Efficacy of Pesticide Application, research report, Dept. of Chemical Engineering, University of New Brunswick, Fredericton, N.B.

Picot, J.J.C., M.W. van Vliet, and N.J. Payne (1989): Droplet size characteristics for insecticide and herbicide spray atomizers, Can. J. Chem. Eng., **67**, pp 752-761.

Picot, J.J.C., M.W. van Vliet, N.J. Payne, and D.D. Kristmanson (1990): Characterization of aerial spray nozzles with laser light-scattering and imaging probes and flash photography, Liquid Particle Size Measurement Techniques, **2**, ASTM STP 1083, E.D. Hirleman, W.D. Bachalo, and P.G. Felton, eds. ASTM, Philadelphia, PA, pp 142-150.

Prandtl, H., and O.G. Tietjens (1957): Fundamentals of Hydro and Aeromechanics, Dover Publications Inc. (translation of the 1934 original in German).

Quantick, H.R. (1985a): Aviation in Crop Protection, Pollution and Insect Control, Collins, London, UK.

Quantick, H.R. (1985b):Handbook for Agricultural Pilots, 4th edition, Collins, London , UK.

Skyler, P.J., and J.W. Barry (1991): Final report- Compendium of Drop Size Spectra Compiled from Wind Tunnel Tests, FPM 90-9, USDA Forest Service Forest Pest Management, Davis, CA.

Tennankore, K., J.J.C. Picot, B. Chitrangad, and D.D. Kristmanson (1980): Aircaft vortex studies in forest aerial spraying, Trans. Of ASAE, **23**, no.5, pp1076-1080.

Tomney, T.D. (1978): The Characteristics of Drop Size Distributions Generated by Aerial Spray Atomizers, M.Sc. Thesis, Dept. of Chemical Engineering, University of New Brunswick, Fredericton, N.B.

van de Hulst (1981): Light Scattering by Small Particles, Dover Pub. Inc., New York.

van Vliet, M.W., and J.J.C. Picot (1989): Calibrating light imaging and light scattering probes for atomizer spray characterization, Pesticide Formulations and Application Systems, **8**, ASTM STP 980, D.A. Hovde and G.B. Beestman, eds., ASTM, Philadelphia, PA, pp 172-192.

Wallace, David (1988): Aircraft Vortex Effects in Forestry Aerial Spraying, Ph.D. thesis, University of New Brunswick Dept. of Chemical Engineering, Fredericton, N.B.

Wallace, D.J., J.J.C. Picot, and T.J. Chapman (1995): A numerical model for forestry aerial spraying, Agric. and Forest Met., **76**, pp 19-40.

Wickens, R.H. (1977): Aircraft Wake Vortices and Their Effects on Aerial Spray Deposition, National Research Council Canada Report LTR-LA-219, Ottawa Canada.

Yates, W.E., E. Cowden, and N.B. Akesson (1985): Drop spectra from nozzles in high-speed airstreams, Trans. ASAE, **28**, pp 405-410.

Young, B.W., and W.D. Bachalo (1988): The direct comparison of three "in-flight" droplet sizing techniques for pesticide spray research, Optical Particle Sizing, Theory and Practice, G. Gouesbet and G.Grehan eds. Plenum Press, New York, pp 483-497.

CHAPTER 5

RESEARCH FIELD TRIALS

5.1 Introduction

In a study assessing priorities for research in forestry insecticide application technology, Campbell and Howard (1994) surveyed users, applicators, researchers, and regulators in Canada and the United States. Respondents were asked to rank up to 52 research topics in order of importance. The top five topics selected were the following:

1. Increase the residual efficacy of Bt and viruses.
2. Determine how to develop public support for aerial application of pesticides in forestry.
3. Determine appropriate, scientifically-based buffer zone widths required to prevent significant environmental impact in sensitive areas.
4. Develop techniques for rapid and accurate determination of Bt and virus tank potency.
5. Develop a practical deposit assessment technique for operational Bt and virus applications.

Another topic prominent among the top ten choices was verification of spray models for insecticide applications to at least 1 km downwind using definitive field experiments. Among the top priorities of the (U.S.) National Steering Committee for Management of Seed, Cone, and Regeneration Insects [7th report, Barry (1996)], are the conducting of pilot tests for new insecticides and the investigating of improved efficiency and efficacy in pesticide application methods. The principal objective of the (Can.) Spray Efficacy Research Group as expressed in the annual report of the Integrated Forest Pest Management Initiative [Caldwell (1995)] is the improvement of the technology in the use of pesticide control products.

Field trials are needed to satisfy several of the objectives defined in these surveys. In this chapter we will endeavour to provide some guidance in the design and execution of successful field trials. In our experience, these trials are usually complex and require careful planning. In order to cover the range of expertise needed to measure key parameters the services of a number of agencies may have to be called on. Co-ordination and scheduling of several groups, subject to the vagaries of weather, is essential. The discussion which follows is based largely upon our experiences in the Dunphy trials, mentioned in Chapter 1. The purpose of these trials was to measure foliage and ground deposit and airborne drift off the experimental plot and to relate them to operational and meteorological factors.

5.2 Measurements and Analytical Methods

Most often, insecticide or herbicide is intended to be applied to foliage. In the first case this is to protect the vegetation against insect defoliators, and in the second, to eradicate unwanted plant species. Application technology research has focussed on the physical processes involved in placing pesticide particles or droplets on foliage. There are usually practical objections to using the actual insecticide or herbicide in repeated trials on the same site so that much work has been done using pesticide simulants with physicochemical properties closely resembling those of the actual formulation. Chapter 2 has dealt with the problem of foliage density and vertical distribution and particle impaction on foliage. The need to use foliage simulators is addressed as well. These are used when the number of separate tests in a trial precludes the use of actual foliage because of the problem of getting accurate "background" deposit levels before each test as these levels continue to build up from one test to the next. Subtraction of before-test from after-test deposits when actual foliage is used often yields negative deposit results because of the normal temporal and spatial variation in deposit density on foliage.

Foliage deposit measurement using simulators

The method of using foliage simulators to estimate foliage deposition will be outlined. The design and development of foliage simulators have been described in Chapter 2. Although wind tunnel tests have been done to relate collection efficiency of this design of simulator to droplet size, airspeed, and simulator orientation [section **2.4** and Picot et al.(1993)], these results cannot be used directly in field testing. At the simulator location, the droplet size spectrum is not well-defined nor is the wind velocity. Instead, the procedure followed has been to expose fresh foliage alongside simulators during a test and to compare the two deposits. These tests give a proportionality factor between foliage and simulator deposits which can be used to estimate total foliage deposit at a station. Field experience has shown this procedure can be justified if the forest stand is reasonably uniform and if the droplet size spectrum does not change markedly in the downstream movement of the droplet cloud (the usual case in the first 500m in ULV spraying). In the Dunphy trials, this direct method of calibration was used and a single calibration factor (Table 2.4) for the dominant species was used for all sampling stations. A similar approach has been used for broadleaf foliage (section **2.4**).

The simulators should be placed in a three dimensional pattern which:

a: is consistent with the relationship between foliage density and height in the forest, and
b: identifies rapid changes in the forest profile, i.e. requiring more simulators in the near field.

The Dunphy, N.B. test site layout is shown in Figure 5.1 and the downwind locations of trees used to set up the simulator lanyards (depicted on Figure 2.8) are given in Table 5.1 [van Vliet (1987)]. The foliage simulators themselves were attached to the lanyards at points equivalent to the centroids of the top, middle, and lower thirds of the cumulative foliage distribution curves for the test plot (see Figure 2.1). The sample trees were selected to provide three lines of sample sites parallel to the 245° direction (Figure 5.1; note the shift in the direction of the spray line between the initial development of the site in 1984 and later extensive testing) and with a 30 meter lateral spacing between lines B,C, and D (Table 5.1). For each downwind position, 18 simulators were evenly distributed between the nine positions presented by three trees at three heights. The simulators were collected together by downwind location, washed with hexane, and analyzed by gas chromatography for the tracers tris-2-ethylhexylphosphate and triphenyl phosphate present at about 2% to 3% in the pesticide simulant used in the trial. Typically, the hexane solutions were concentrated by evaporation at room temperature and analyzed on a Hewlett Packard 5890 gas chromatograph equipped with a nitrogen/phosphorus detector and autosampler, and using a 30 m, DB 608 megabore column with helium as the injector gas at ca. 15 ml/minute at an oven temperature of 250°C [Riley (1992)]. Analysis of exposed foliage for calibration of the simulators was done by chopping and then macerating a 10 g sub-sample with 100 to 200 ml of a 5% acetone/hexane solvent in a blender. Samples were allowed to settle and a sample of the extract was used for the GC analysis. The amount of tracer present in an aliquot was then used to calculate the total amount on foliage simulators at the sample location. Using the factors in Table 2.4, this was used to calculate tracer on a gram of foliage or drops per needle. The measured foliage density for the test site permits the calculation of tracer on foliage per unit area of ground for the location. With a series of samplers distributed downwind, the total foliage deposit can be calculated and related to the aircraft emission per meter of flight path. This can be done by visualizing a 90-meter wide stripe downwind of the flight line sampled in a regular array, having been sprayed over the 90 meter long flight path with a constant and metered application rate of a simulant containing a measured amount of tracer. This method can be used in either insecticide or herbicide application situations.

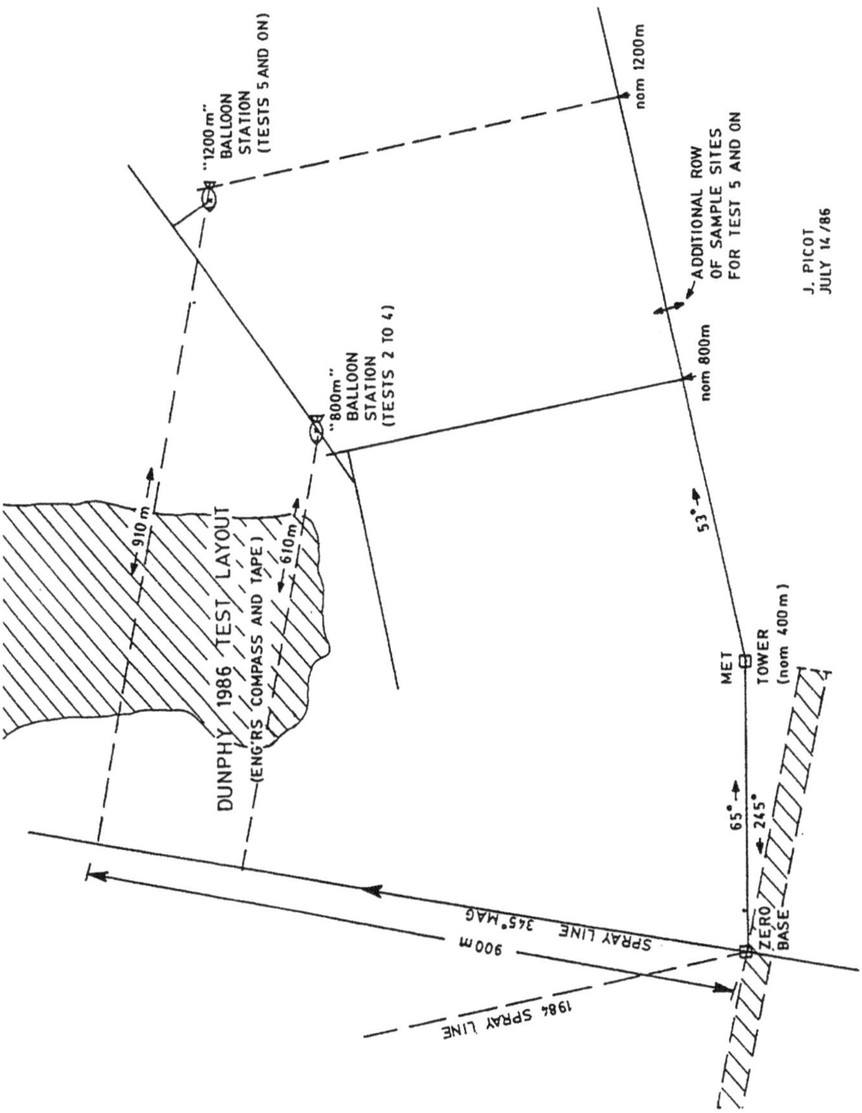

Figure 5.1 Sketch of the Dunphy, N.B. test site located at 46,580N, 65,830W. Drawing to scale

TABLE 5.1 Foliage sample station details, Dunphy test site

Nominal and Actual Distances from Spray Line

Nominal Position		Actual Position (m from spray line)	Downwind Distance to be used	Nominal Position		Actual Position (m from spray line)	Downwind Distance to be used
20	B	25	19.8	100	B	104.5	86.5
	C	14.5			C	90	
	D	-3*			D	65	
30	B	31	27.0	110	B	112	114.5
	C	23			C	117	
40	D	2*		130	B	131	117.0
	B	38	28.0		C	108	
	C	32			D	112	
50	D	14		150	B	142	150.2
	B	51.5	40.0		C	150	
	C	42.5			D	158.5	
60	D	26		200	B	1945	193.5
	B	60.5	46.5		C	197	
	C	44			D	189	
70	D	35		250	B	247	234.0
	B	68	54.3		C	235	
	C	57			D	220	
80	D	38		300	A	301.5	294.0
	B	75	61.3		B	290.5	
	C	67			C	290	
90	D	42		350	A	348	334.2
	B	90.5	73.7		B	335	
	C	77			C	319.5	
	D	53.5		400	B	374	363.0
					C	353	
					D	362	

* not included in average downwind distance

The calculation can be summarized as follows, for the mass of tracer deposited on foliage in the downwind interval between two foliage deposit sampling sites, for each tree species present:

$$\text{Interval Fol. Dep.} = \{\text{Ave. Fol. Dep. (eq'n 2.4.2)}\} \times \text{SI} \times \text{P} \qquad (5.2.1)$$

Here, Ave. Fol. Dep. Is calculated from eq'n 2.4.2 from upwind and downwind

Fol. Dep. For the interval, and

SI =	the distance (measured parallel to wind direction) between upwind and downwind sample location for each sampling interval, m, and
P =	flight distance across sampling plot (90 m in above example)

This calculation gives foliage deposit in grams for the sample area between sampling intervals. To get the fraction of aircraft-emitted tracer collected by foliage, the result of equation 5.2.1 is divided by the mass of tracer emitted by the aircraft as it crosses the sample plot:

$$\text{Tracer emitted} = \frac{E}{V} \times P \qquad (5.2.2)$$

Here,	E =	aircraft tracer emission rate, g/s
	V =	aircraft ground speed across sample plot, m/s
	P =	flight distance across sample plot, m

"Kromekote" paper strips can be attached to the simulator tines to permit droplet size monitoring when the spread factor of the simulant droplets on the paper is known [Kristmanson et al. (1987)].

Foliage Deposit Using Fresh Foliage Samples

In a paper reviewing spray trial measuring techniques, Riley (1995) has pointed out that the vegetation itself is the preferred sampling medium for canopy deposit, except for the reservation indicated above. An alternative to the simulator is to cover up foliage which is to be used in subsequent tests. Colorimetry or fluorometry will give complete recovery of deposits of tracer on simulators but dye recoveries can be poor from natural vegetation.

Droplets on foliage can be sized by microscopic methods but there are two inherent difficulties, particularly if the information desired is the total deposit. First, much of the mass of the droplets in a typical emission spectrum is found in a relatively few large droplets. The most frequently seen droplets do not contain much of the mass. Typical spraying of conifers yields very few droplets larger than 70 microns on foliage. This fact was pointed out by Himel and Moore (1967) and Barry and Ekblad (1978) and can be confirmed by considering the fact that for a Foray 76 Bt product sprayed in a Micronair AU4000 at a rate of 5.0 l/min, and at a rotational speed of 10,000 rpm, and an aircraft air speed of 50 m/s, 99% of all generated droplets are smaller than 73 microns (see Apppendix II). This is a typical result, and in this case, the $D_{v.5}$ was about 70 microns.

A second difficulty arises because of the ambiguity of the relationship between the size of the stain or spread-out droplet on the foliar surface. What is measured by microscopic examination may not be readily converted into a number spectrum of collected droplets. Deposits of Bacillus thuringiensis insecticide have been evaluated in this way but mass calculations were found to be difficult [Riley et al. (1992)]. Kettela (1995) has also concluded that the microscopic method does not give an "absolute" measure of spray deposit for Bt formulations partly because of the variability of behaviour of dyes in the different formulations.

Canopy deposition has been measured in USDA Forest Service spray trials using 2.5 inch diameter soft drink cans suspended on and between the trees in a test plot. Mylar wrapping or Kromekote card wrapping was used to collect droplets for tracer analysis and droplet sizing [Barry et al. (1982)]. This technique of assessing the "impactable" droplets subsequently evolved to the use of static Rotorods (see below) with 2 mm square cross-section U-shaped wire collectors [Barry (1991), Barry et al.(1992)], which have impaction characteristics closer to those of conifer needles than do the cans.

In other work, Kirk et al. (1989) report on the use of soft-drink straws to measure the deposit of herbicide on yellow foxtail, and Raske et al (1989) report on the use of the UNB simulators mentioned above to assess deposit in a conifer stand. Uk and Courshee (1982) have reviewed the mechanisms of droplet capture on artificial and biological surfaces and discuss the effect of spray parameters on deposit for each type.

Ground deposit

Measurements of ground deposit seem conceptually simpler than foliage deposit measurements, but the complicated structures present in natural ground covers require careful consideration. In a forest in which the tree canopy is much higher than the ground cover, it is logical to incorporate deposits on grasses, ferns, etc. in the category "ground deposit". On the other hand, the horizontal wind speed is very low or zero at ground level, and an assumption that the droplet velocity is essentially vertical means that any flat, horizontal surface from which deposited droplets can be adequately removed provides a good sample. In the Dunphy insecticide tests [Picot et al. (op. cit.)], rectangular aluminum plates (20 cm x 40 cm) were placed on the ground at each sampling station. These were placed so that they were not sheltered by small plants. These were collected after each trial, washed with solvent, and the solvent analyzed for the tracers as for the foliage samplers mentioned above. Ground deposit for each sample interval is obtained by a similar calculation to equation 5.2.1:

$$\text{Ground deposit} = \text{Ave. sampler dep.} \times \text{SI/GSA} \times \text{P} \qquad (5.2.3)$$

where GSA is ground sampler area.The assumption of negligible horizontal velocity means that the deposit values are also correct for ground level vegetation.

USDA Forest Service practice is similar in that Kromekote cards and Mylar sheet are placed flat on plywoood panels at ground level [Barry et al. (1991)]. A similar approach has been used by Riley et al.(1989) and Riley and Wiesner (1990) in agricultural and orchard spray studies. In a forest herbicide study, Shewchuck et al. (1991) placed Petri dishes and and flat paper and film sheets on a horizontal plywood panel about 1 meter above ground level. Spray material was collected and analyzed and the result converted to total deposit. On the other hand, herbicide spray testing done by SERG requires that ground deposit samplers be deployed on flat boards lying on the ground, in cases where foliage deposit is measured separately [Riley (1992)].

Suspended droplet concentration and spectra

The measurement of off-target drift from a spraying operation implies the measurement of air flow velocity away from the spray zone and the mass concentration of particles suspended in the air. The particle (droplet) size distribution is also of interest. Hesketh (1972) reported an estimate of the National Air Pollution Control Administration that there were 14,685 air quality sample collecting and monitoring devices available. For suspended droplets, sampling is done by aspirating air into a collecting system for either total mass measurement or for size segregation before measurement of mass (as in the cascade impactor), or by moving a collecting surface at appropriate velocity through the air (as with the Rotorod sampler [Edmonds (1972)] shown on Figure 5.2). In the use of air aspirating samplers, it must be realized that the design and size of the entrance tube determine the upper limit of accurate droplet sampling. Large droplets are effectively removed from an air stream which is forced to move through a high-curvature path into a sampling tube. The design considerations for this type of sampler and specifications of some commercial devices are dealt with in the text edited by Liu (1976) in his Part III on aerosol sampling.

The Rotorod sampler has become the preferred sampler for pesticide droplet drift [McCooeye et al. (1993), Barry(1991), Crabbe and McCooeye (1990 and 1988), Riley and Wiesner (1990), Riley et al. (1989)]. The collection efficiencies for the rods can be estimated using the May and Clifford correlation for a ribbon collector (eq. 2.3.2).Using the U-rod collector (1.59 mm wide square cross-section collector with arms 6 cm high and 8 cm apart) rotating at 2400 rpm, the collection efficiency vs Stokes number correlation equation 2.3.2 (the May and Clifford data for a ribbon collector) gives a collection efficiency of 67% for 10-micron droplets of density unity, using Figure 2.4 to estimate settling velocity, and an efficiency of 81% for 15-micron droplets. With the H-rod (0.38 mm wide square cross-section collector with arms 6 cm high and 12 cms apart) the collection efficiency for 10-micron droplets is calculated to be 92%. The volume sampled by the U-rod is 115 litres per minute and that by the H-rod is 41.3 litres per minute, based on material collected on the parallel arms. Kristmanson et al.

(1987) have shown that Kromekote paper strips can be glued to the Rotorod surfaces to obtain droplet size distribution. When these collectors are used alongside the conventional collectors, it is possible to calculate a global collection efficiency for the conventional collectors when particles smaller than 15 and 10 microns are present, respectively, for the U and H collectors. This can be done for each size class as follows:

$$\text{Actual mass collected} = \Sigma_i \frac{\pi\rho}{6} D_i^3 n_i$$

$$\text{Corrected mass collected} = \Sigma_i \frac{\pi\rho}{6} \frac{D_i^3}{\eta_i (D)} n_i \qquad (5.2.4)$$

Here, n_i = droplets per unit time in size class i (measured)
D_i = diameter (mean) of size class i
η (D) = collection efficiency of Rotorod for droplets of diameter D (calculated, equation 2.3.2)

The ratio of actual and corrected collected mass is found from the above values and divided into the actual mass collected on conventional Rotorods to give corrected mass sampled. Note that a lower limit of droplet collection efficiency of very small droplets can be assumed as equal to droplet diameter divided by collector surface width. The droplet splattering which can occur at droplet sizes greater than 60 microns in diameter and the overloading of the Kromekote strips at high concentrations will vitiate this technique. May et al (1976) measured sampling efficiencies of 85.7, 85.2, and 93.1% for droplets of 50, 30, and 20 μm on the U-rods and state that the lower efficiency for larger droplets is probably due to drop shatter from droplets striking the edges of the collecting surfaces.

McCooeye et al (op. cit.) have used vertical arrays of modified Rotorods (with 0.6 mm wide collectors) carried by blimps to heights up to 300 meters at the 600m downwind position in a field trial. In earlier work, this group (the NRC Unsteady Aerodynamics Laboratory in Ottawa) used dosimeters developed expressly for suspended droplet sampling and analysis [Crabbe and McCooeye (1990)]. These are described on Figure 5.3.

Meteorology

A variety of instrumentation has been used in research spray trials, in keeping with the different objectives of the studies. A study plan [Barry 1991, op. cit.] for off-site movement of pesticide in complex terrain called for three EMCOT [Ekblad et al. (1990)] solar powered portable weather stations providing the following weather data at 2 meter and 6.5 meter heights:

Wind speed
Wind direction

Relative humidity
Cloud cover

The data collected permitted calculation of mean velocity, temperature, and relative humidity profiles as well as turbulence [Barry et al., op. cit.].

A trial in a southern pine seed orchard [Barry et al. (1982)] was instrumented with MRI Vectorvanes (Figure 5.4) mounted at 2 meters and 16 meters to measure azimuth and vertical wind direction and speed and a tethersonde was used to get wind direction, wind speed, temperature, and relative humidity measurements up to 70 meter heights.

The 1986 Dunphy study was instrumented with six levels of MRI Vectorvanes on a 63 meter mast, and these were fitted with microbead thermistors to give temperatures; relative humidity was measured with a capacitance probe (Humicap, from Vaisala, Inc. Woburn, MA) [Crabbe and McCooeye (1991)]. In addition, minisonde and tethersonde measurements were made above tower

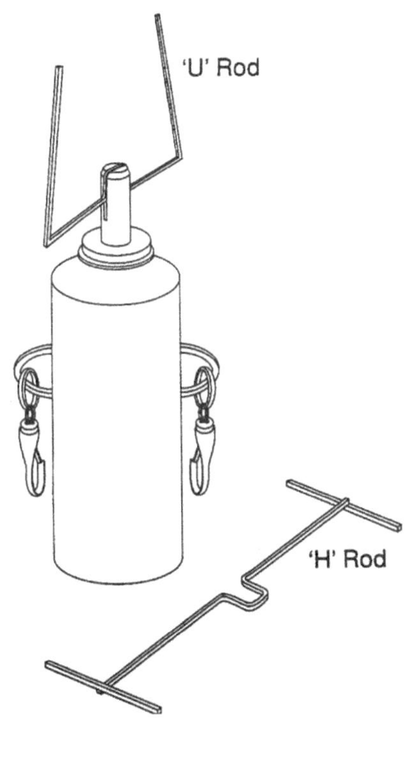

Figure 5.2 The Rotorod sampler, Ted Brown Associates, Los Altos Hills, CA. Operates at 2400 rpm at 12 volts d.c. Sketch copied from McCooeye et al. (1993) by kind permission of the author.

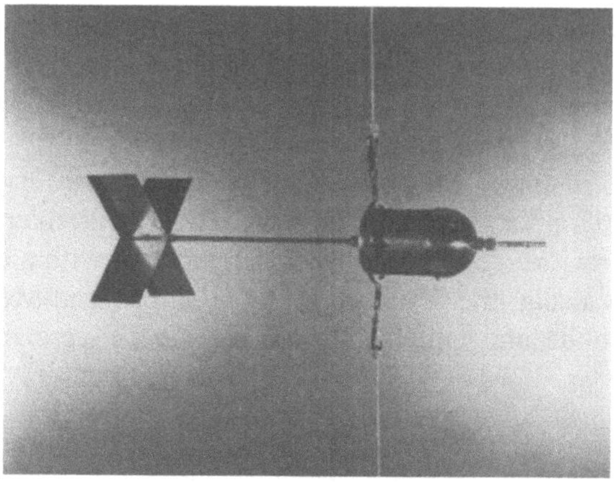

Figure 5.3 Photograph of the NRC dosimeter. Sample tubes are glass and are 6.2 mm o.d. by 7.5 cm long, packed with 0.5 g of fine Pt mesh plug 1.5 cm in length [Crabbe et al. (1980)]. Copied by kind permission of the author.

height, and a pair of shielded MRI transistor thermometers were used to measure temperature difference between tree height and tower top.

In a "conifer release" herbicide study, Riley (op. cit.) reports the use of a 18.8 meter tower instrumented for wind speed and direction at 2.2, 4.1, 8.0,and 16.4 meters, and with a MRI Vectorvane at the top for turbulence measurement. Temperature sensors were positioned at 2.2, 8.0, and 16.4 meters with relative humidity measured at 8.0 meters. Digital data logging was done with a Campbell Scientific 21x data logger and five-minute averages were calculated.

Minimum Length of Spray Line

Flight paths over spray test plots are usually meant to simulate "line source" emission and must be long enough so that lateral dispersion from the beginning and end of the line can be ignored. The critical length, assuming flight at right angles to wind direction, can be assessed by using plume spread parameters from Gaussian plume modelling [e.g. Hesketh (1972)]. Assuming unstable (maximum dispersion) conditions, a horizontal dispersion standard deviation can be found graphically from that text to be 200 m at a downwind distance of 1000 m from the source. Therefore, at a lateral distance of 200m from a point source emission and 1000m downwind, the concentration is 67% of the value at the plume center-line. This can be interpreted as the zone significantly affected by the beginning and end

Figure 5.4 MRI Vectorvane providing analog readout signals for wind speed, azimuth and its standard deviation, and vertical angle and its standard deviation. It is shown here with elevation angle calibrator (normally absent) locked to the vertical post.From Meteorology Research, Inc., Alhambra, CA

of the spray line. Conservatively, at the downwind extremity of the test plot, there should be at least a 400-meter lateral distance between the end of the spray line and the edge of the test plot. In the ideal case of flight at right angles to wind direction, this means that for a 90-meter wide test plot, the flight line with spray booms "on" should start at a point 445 m up-track of the plot center-line, and end no sooner than 445 m down-track. Since this is the ideal and is extremely restrictive, additional up-track and down-track distance is required to allow a realistic range of angular deviation from right-angle winds over a test plot. When angular deviations are large, it becomes necessary to estimate the affect of lateral dispersion on the results, even when the spray line is long enough to avoid end effects. Alternatively, it is possible to design a test plot which will accommodate a range of wind directions by permitting relocation of samplers. This is expensive and not often practical.

The above-mentioned restriction is less severe in stable conditions. The horizontal standard devation is 35 m rather than 200 m as in the unstable case.

5.3 The Mass Balance Test as a Measure of Accuracy in Field Trials

The non-volatile tracer material released by a spray aircraft over a test plot will be found 1) deposited on foliage in the immediate aircraft swath and downwind, and 2) deposited on the ground in the same region, and 3)carried downwind of the deposit sampling area. This assumes that settling of droplets in the deposit sampling area is finished. Foliage and ground deposit values are obtained from equations 5.2.1 and 5.2.3 for the deposit sample zone, giving integral foliage and ground deposit for the zone. At the downwind extremity of the sample zone, vertical traverses of droplet concentration and windspeed are required to permit calculation of off-site drift, as follows [Crabbe et al. (1980)]:

$$\text{Downwind droplet transport} = \Sigma_i(C_i \cdot u_i \cdot \Delta z_i \cdot P) \cdot \Delta t \qquad (5.2.5)$$

Here, C_i = mass of tracer collected in the sampling period divided by total air volume sampled in that period, for the vertical sampling interval Δz_i, g/m^3

Δt = sampling period, s

u_i = mean horizontal wind velocity for interval Δz_i for the sampler time period and position, m/s

Δz_i = step size in vertical sampling grid

P = Flight distance across sampling plot, m

The dampling period must be long enough for all drifting droplets to cross the downwind boundary of the test plot.

Provided that droplet cloud sampling reaches the top of the droplet cloud, the sum of all three items above will give a value equal to the value for tracer emission given by equation 5.2.2. This constitutes a powerful test of the overall accuracy of the measurements done, i.e., the "mass balance test". The mass balance closure is the sum of the three terms above divided by the emitted tracer mass from eq'n 5.2.2. This ratio has also been called "pesticide recovery" and is often expressed as a per cent value.

Crabbe and McCooeye (1990) report that the average mass balance closure for the Dunphy trials was 77% for the 1984 series and 87% for the 1986 series with the improvement dependent on better sampling technology. Picot et al. (1993) carried out a regression analysis for the Dunphy trials where total deposit was regressed against a) vapor pressure driving force (i.e. evaporability of water in the formulation), b) aircraft height above ground, c) wind direction deviation from perpendicular to flight line, d) vortex sink rate, e) windspeed, and f) Monin-Obukhov length. Only the aircraft height showed a statistically significant relationship for the 22 observations used. The Crabbe and McCooeye (op. cit.)

paper includes the results of later tests in Northern Ontario and concludes that neutral to unstable conditions reduce drift relative to stable conditions. In these tests, mass balance closure averaged 75% for the stable cases and 51% for the unstable cases. There was some difficulty in measuring deposit on the mixed-type vegetation in the Northern Ontario test site.

The McCooeye et al.(1993) report on a study of drift from upwind wing and downwind wing for an aircraft flying at right angles to wind direction concluded that on average, the upwind wing deposit was 1.5 times greater than the downwind deposit for an unforested test plot out to 200 m downwind. Drift results beyond 200 m showed an 18% lower drift from the upwind wing. Mass balance closures at the 200 m position averaged 94% for the upper wing and 89% for the lower wing, with individual values for the nine tests ranging from 51% to 151%.

In all tests, low values obtained for mass balance closure should elicit caution when reaching conclusions. Sources of probable error can be listed as follows in order of diminishing importance:

1. Deposit and drift sampling error (including wind speed error)
2. Aircraft pump flow rate measurement error
3. Aircraft tank mix sampling error
4. Aircraft speed measurement error
5. Spray line location error
6. Sample analysis error

We conclude that experimental design should incorporate a requirement for measuring all the variables needed for calculating mass balance closure. In addition, we have found that the ultimate tool for evaluating the effects of individual spraying parameters is a validated simulation model, rather than a field trial with a complicated experimental program. Once a new spraying modification has been identified as beneficial by testing with the model, the critical well-designed field trial can be done for proving the modification.

5.4 Experimental Design, Management, and Costs

It is possible, in theory, to set up a statistical experimental design to evaluate the effect of experimental variables systematically, but, in practice, the execution of the design is difficult, unless it is very simple. The major difficulty arises from meteorological factors, important in the analysis of the results and impossible to control. If experiments are conducted in which one variable, say aircraft height, is being varied and it is intended that meteorological conditions remain constant, at least approximately, it is usually difficult to get a second trial on a given day.

The development of stable to unstable conditions in the morning (as discussed in Chapter 2) and the reverse in the late afternoon requires that two trials must be done over a short time interval. In practice it has been found that a fast turnaround between trials is not easy because of the time required for sample collection and the deployment of drift samplers.

In the authors' experience, field trials should be governed by the following rules:

1. Identify exactly what the test objectives will be.
2. Select the test site and design the test plot to best meet the objectives.
3. Identify test equipment and methods.
4. Describe the series of experiments with the values of test parameters desired.
5. Identify the necessary equipment and the necessary services.
6. Set up cost estimates and organize funding in advance.
7. Secure the necessary authorizations for spray application in the environment
8. Assemble the teams and organize the work plan with intensive group planning sessions where all participants are briefed and have an opportunity for input to operational logistics planning.
9. Carry out the testing, but remain sensitive to the need to modify plans in view of unforeseen events.

Item 9. above reflects the fact that there is usually a limited time available for a field trial, and that once a large part of the time has elapsed, it is often necessary to review the objectives to see if the test plan is still achievable or if modifications are called for. Any single valid test is preferable to a completely aborted series of tests. Contingency planning in the military sense is useful.

The USDA Forest Service preliminary plan for a study of off-site movement of Bt spray applied in complex forested terrain [Barry (1991), 43 pages] summarizes the organizational effort under the following headings:

PREFACE ....
INTRODUCTION ....
OBJECTIVE ....
SCOPE ....
METHODS ....
    Application ....

Cont'd....

A characteristic of many spray trials is the involvement of many different jurisdictions and agencies. For example, SERG trials typically involved Forest Protection Ltd. (aircraft and spray equipment, field accommodations for personnel), NRC Unsteady Aerodynamics Laboratory (drift sampling and low-level meteorology), Atmospheric Environment Services (turbulence measurements, lidar mapping of spray cloud) , UNB Forestry (upper level minisonde measurements), UNB Dept. of Chemical Engineering (atomizer characteristics, foliage characterization and deposit measurement, simulation model validation), NB Research and Productivity Council (chemical assay and project management), and Forestry Canada (insect sampling and behaviour, project management). Successful trials were the result of good planning and clear communications before and during the trials. The large projects required a field coordinator and frequent plenary briefing sessions during the trials.

Field trial costs have the following components:

1. Salaries of professionals involved (managers and scientists)
2. Costs of aircraft operations
3. Costs of analytical services
4. Purchase or rental cost of equipment
5. Rental and operating costs for vehicles
6. Accommodations, food, and travel costs

7. Salaries for casual workers needed for site preparation and security, and sample deployment and collection
8. Costs of characterization of atomizers
9. Costs for report preparation

For the SERG drift field trials at Dunphy, N.B. in June 1984 (ten-day experimental program, one Grumman TBM aircraft) and the Dunphy 1986 spray trials in July (19-day experimental program with one TBM and one Cessna 188 aircraft), the total costs amounted to C$ 278,000, exclusive of aircraft costs which were an in-kind contribution from Forest Protection Ltd [Davies (1996)]. Approximately $59,000 went for chemical analysis costs, and $102,000 went for site preparation (including meteorological tower construction), food and accommodation, formulation preparation and aircraft support, and site security. Of the total scientific personnel, 10 were professionals, 5 were technicians, and 5 were university students. Salaries of eight of the professionals were borne by their agencies. All personnel were housed and fed at the Forest Protection Ltd. and New Brunswick Natural Resources camp facilities at Dunphy airstrip (five miles from the test site in central New Brunswick).

References

Barry, J.W. (1996): 7th Report: National Steering Committee for Management of Seed, Cone, and Regeneration Insects, USDA Forest Service Forest Health Protection, Davis CA.

Barry, J.W. (1992): Predicting Spray Drift in Complex Terrain, paper #921085, 1992 International summer meeting, Amer. Soc. of Ag.Eng'rs, Charlotte, N.C.

Barry, J.W. (1991): Study plan: Off-Site Movement of Bt Spray Applied to Complex Forested Terrain, USDA Forest Service Forest Pest Management, Davis, CA.

Barry, J.W., P.A. Kenney, L.Barber, R.B. Ekblad, R.J. Dumbauld, J.E. Rafferty, H.W. Flake, and N.A. Overgaard (1982): Aerial Application to Southern Pine Seed Orchards, Data report of the Withlacoochee trials, USDA Forest Service, Southern Region Forest Pest Management, Asheville, N.C.

Barry, J.W. and R.B. Ekblad (1978):Deposition of insecticide drops on coniferous foliage, Trans ASAE, **21**, pp 438-441.

Caldwell, E. (1995):Integrated Forest Pest Management Annual Report 1994-1995, Natural Resources Canada, Canadian Forest Service, Sault Ste. Marie, Ont.

Campbell, R.A. and C.A. Howard (1994): Priorities for forestry insecticide application technology research, J. Envir'l Sci. and Health Part B: Pesticides, Food Contaminants, and Agricultural Wastes, **29**, pp 561-619.

Crabbe, R.S., and McCooeye, M.A. (1988): Field Procedure for Measurement of Flightline Offset in Forestry Spraying, National Aeronautical Establishment Aeronautical Note #NAE-AN-51 (NRC # 28790), National Research Council, Ottawa.

Crabbe, R.S., and McCooeye, M.A.(1990): Wind drift from ultra-low volume aerial insecticide applications in stable and turbulent conditions, Proc. ACAFA Symp. on Industries Future Needs, Winnipeg, Man. Pp 158-168.

Crabbe, R.S., and M.A. McCooeye (1991): Drift Cloud Mass and Meteorology in the Dunphy 1984/86 Field studies, unpublished, National Research Council, Ottawa.

Crabbe, R.S., L. Elias, M. Krzymien, and S. Davie (1980): New Brunswick Forestry Spray Operations: Field Study of the Effect of Atmospheric Stability on Long-Range Pesticide Drift, National Aeronautical Establishment Report LTR-UA-52, NRC, Ottawa.

Davies, D. (1996): personal communication, Managing Director, Forest Protection Ltd.

Edmonds, R.L. (1972): Collection efficiency of Rotorod samplers for sampling fungus spores in the atmosphere, Plant Disease Reporter, **56**, pp 704-708.

Ekblad, R., K. Windall, and B. Thompson (1990): EMCOT Weather Station, USDA Forest Service, Missoula Technology Development Center, Missoula, MT.

Hesketh, H.E. (1972): Understanding and Controlling Air Pollution, Ann Arbor Science Publishers, Ann Arbor MI.

Himel, C.M., and A.D. Moore (1967): Spruce budworm mortality as a function of aerial spray droplet size, Science, **156,** pp 1250-1251.

Kettela, E.G. (1995): Evaluation of Quality and Quantity of Bt Spray Deposit on White Spruce, report to SERG, available at Can Inst. For Scientific and Technical Info., Ottawa.

Kirk, I.W., L.E. Bode, L.F. Bouse, R.A.Stermer, and J.B. Carlton (1989): Deposition efficiency from aerial application of postemergence herbicides, Pesticide Formulations and Application Systems, International Aspects, 9th Vol., ASTM STP 1036, J.L. Hazen and D.A. Hovde eds. ASTM, Philadelphia, PA. pp 211-232.

Kristmanson, D.D., J.J.C. Picot, S. Van Vliet, and G.W. Henderson (1987): Measuring foliar deposits from aerial spraying of pesticides, Proc. Symp. On Aerial Application of Pesticides in Forestry, National Research Council # 29197, NRC, Ottawa, pp 269-273.

Liu, B.Y.H. (1976): Fine Particles, Academic Press, New York.

May, K.R., N.P. Pomeroy and S. Hibbs (1976): Sampling techniques for large windborne particles, J. Aerosol Sci., **7**, pp 53-62.

McCooeye, M.A., R.S. Crabbe, R.E. Mickle, A. Robinson, E.B. Stimson, J.A. Arnold, and D.G. Alward (1993): Strategy for Reducing Drift of Aerially Applied Pesticides, Institute for Environmental Chemistry, National Research Council Canada, Ottawa.

Picot, J.J.C., D.D. Kristmanson, R.E. Mickle, R.B.B. Dickison, C.M. Riley, and C.J. Wiesner (1993): Measurements of folial and ground deposits in forestry aerial spraying, Trans. ASAE, **36**, pp 1013-1024.

Raske, A.G., K.M.S. Sundaram, A. Sundaram, and R.J. West (1989): Fenitrothion deposits on simulated and live fir foliage following aerial spraying of two formulations, Pesticide Formulations and Application Systems: International Aspects, 9th Vol., ASTM STP 1036, J.L. Hazen and D.A. Hovde eds., ASTM, Philadelphia PA, pp 233-243.

Riley, C.M. (1992): Continuation of Studies to Examine the Effects of Aircraft Application Parameters on Deposit and Drift of Forestry Herbicides, Report #C/92/443 (a), NB Research and Productivity Council, Fredericton, N.B.

Riley, C.M. (1995): Techniques for the measurement of pesticide drift, Proc. National Conf. on Pest. Applic. Technology, Ridgetown College of Agricultural Technology, Ridgetown, Ont., pp 45-51.

Riley, C.M., C.J. Wiesner, and W.R. Ernst (1989): Off-target deposit and drift of aerially applied agricultural sprays, Pesticide Sci. **26**, pp 159-166.

Riley, C.M., and C.J. Wiesner (1990) Off-target pesticide losses resulting from the use of an air-assisted orchard sprayer, Pesticide Formulations and Application Systems: 10th Vol. ASTM STP 1078, L.E. Bode, J.L. Hazen, and D.G. Chasin eds., ASTM, Philadelphia, PA, pp 204-225.

Shewchuk, S.R., K. Wallace, and J. Maybank (1991): Spray Drift and Deposit Pattern from a Forest Herbicide Application Saskatchewan Research Council publication E-2310-4-E-90, Saskatoon, Sask.

Uk, S. And R.J.Courshee (1982): Distribution and likely effectiveness of spray deposits within a cotton canopy from fine ULV sprays applied by aircraft, Pesticide Science, **13**, pp 529-536.
van Vliet, S. J. (1987): Aerial Spraying of Forests: Measurements of Ground and Canopy Deposits, B.Sc. Thesis, UNB Dept. Of Chemical Engineering, Fredericton, New Brunswick.

CHAPTER 6

MODELLING SPRAY APPLICATION, DISPERSION, AND DEPOSITION

6.1. Introduction

In a review of the status of forestry pesticide application computational models, Mickle (1987) reported that of the more than twenty models discussed in the literature, four in particular were considered to be superior:

1. AGDISP (*AG*ricultural *DIS*persal model; Bilanin et al (1987), Teske (1986), Bilanin and Teske (1984), Bilanin et al (1978)).
2. Weihs-Atias Model (Weihs and Atias (1987), Atias and Weihs (1985 and 1984)).
3. FSCBG (*F*orestry *S*ervice, *C*ramer, *B*arry, *G*rim; Geary et al (1987), Dumbauld et al (1980)).
4. PKBW (*P*icot, *K*ristmanson, *B*asak-Brown, *W*allace; Picot et al (1987 and 1986)).

More recent publications permit the following up-to-date descriptions of these models:

1. AGDISP predicts the behaviour of spray material released through atomizers into the wake of a spray aircraft, travelling through real atmospheric effects, penetrating and depositing on a canopy, and impacting the ground. The computation is Lagrangian, i.e. droplet trajectories are calculated from emission to deposition. The mean position of droplets is calculated along with the position variance about the mean based on turbulent fluctuations. Solutions are for the near-field effect of the aircraft wake on droplet behaviour. [Bilanin et al (1989), Teske (1990), Teske et al (1991)]. A large number of validation experiments has been done by the USDA Forest Service and model development continues.
2. Weihs-Atias simulates the air flow about a wing and in its wake in the presence of the ground boundary which is described as a wavy surface when a tree canopy is present. Crosswind effects are included, with the aircraft assumed to fly at right angles to the wind direction. Droplet trajectories are then computed using the calculated flow field and the effects of gravity, aerodynamic drag, and evaporation. A number of validation calculations has been done [Weihs and Atias (1987), more recent reference not found].

3. FSCBG is a Gaussian line-source model that takes the near-field results of AGDISP and predicts downwind dispersion and deposition incorporating the effects of evaporation, meteorology, and canopy interaction. This model has been developed by the USDA Forest Service in cooperation with the U.S. Army and has been extensively tested against field data. The model will accept flight lines which are not at right angles to prevailing wind direction, and has been applied to the problem of predicting spray deposition in complex terrain. Parametric sensitivity tests have been done on the model. [Barry et al (1992), Teske and Barry (1993)].There is a FSCBG User Group administered by Continuum Dynamics, Inc., Princeton, New Jersey. A National Spray Model Advisory Committee exists to recommend development and evaluation activities for FSCBG.
4. PKBW was developed by the authors of this text and their graduate students at the University of New Brunswick in Fredericton, N.B., under the sponsorship of the Natural Sciences and Engineering Research Council (Canada), Forest Protection Ltd., Fredericton, N.B., and Environment Canada. The model computes droplet trajectories of the emitted spectrum of droplet sizes under the effects of 1) aircraft vortex, 2) atmospheric turbulence, 3) canopy interaction and deposition, and 4) ground deposition. Evaporation is computed at each trajectory step. Vortex behaviour is computed with the NASA VORTEX WAKE code (see section 4.2). Neutral, stable, and unstable atmospheric conditions are included. The model permits calculations for a multiple-stand situation, e.g. spray application over a young plantation (herbiciding for conifer release) with a mature stand of trees immediately upwind and/or downwind. Deposition calculations in a mixed stand are possible, e.g. a mixed conifer-broadleaf stand. Needle and leaf droplet deposit density and spectrum are calculated. The spray aircraft flight path is assumed to be at right angles to the wind direction. Extensive model validation has been done [Wallace et al (1995)], including comparisons with AGDISP and FSCBG [Riley (1995)]. User workshops have been conducted by Effective Particle Technologies, Ltd., Fredericton, N.B., Canada and coordinated by SERG (Spray Efficacy Research Group, Fredericton, N.B., Canada).

The most important feratures of the AGDISP, FSCBG, and PKBW models are summarized in Appendix III.

In the sections below, we provide details of the PKBW model only, since documentation on AGDISP and FSCBG is readily available from the USDA Forest Service, Forest Pest Management, Davis, CA [e.g. Teske and Barry (1993)]. Comparisons between PKBW and the USDA Forest Service models are discussed.

6.2 Details of the PKBW Model

The PKBW model was conceived as a means of developing a systematic understanding of the mechanisms of forest spraying by aircraft (see Figure 6.1). A number of practical questions are important to applicators. What droplet spectrum is desirable? Should spraying be conducted under stable or unstable atmospheric conditions? Is there an optimum height the aircraft should fly? As mentioned in Chapter 5, field experiments are not a practical way of providing answers to these questions because of the impossibility of controlling some of the important parameters. We chose to develop a mechanistically correct simulation which could be validated by field trials. With confidence in the model, the practical questions of the applicator could be rationally dealt with.

6.1.1 WAKE VELOCITY DISTRIBUTIONS

The WAKE VORTEX code discussed in section 4.2 is used to produce the time-dependent velocity field in the spray aircraft wake immediately after passage of the aircraft. The code assumes a developed vortex immediately behind the wing and a vortex core radius (r_c) of 0.3 m for the small fixed-wing and Grumman TBM aircraft [Wallace (1988)]. The calculated vorticity results for the upwind and downwind vortices are collected at one-second intervals and the total circulation (vortex strength) for each vortex is calculated by integrating vorticity over the complete vortex cross-section. These values are then read into PKBW and used to calculate point velocities and displacements for droplets in the vortex using ideal-flow theory and the method of images [Brashears and Hallock (1974)]. In this way, the vortex decay process is respected without requiring a complete VORTEX WAKE calculation for each PKBW calculation, since a limited number of aircraft-specific vortex calculations over a restricted range of parameters can be used as a data bank for access by PKBW to calculate a large number of site-specific spray situations. After the vortex velocities (for each droplet) have reached 90% of the horizontal mean velocity of the atmospheric boundary layer, the vortex is considered to be decayed and the velocity field is replaced by that of the atmospheric boundary layer.

6.1.2 TRAJECTORY CALCULATIONS

Trajectory calculations are carried out with droplet initial location given by atomizer position on the aircraft, and horizontal and vertical displacements Δx and Δz calculated by assuming that the droplet follows the motion of the air in which it is suspended, with the addition of a gravitational settling velocity. This ignores droplet inertia. The calculation proceeds as follows [Wallace et al (1995)]:

Calculation of the trajectory of a single droplet of known size and properties.

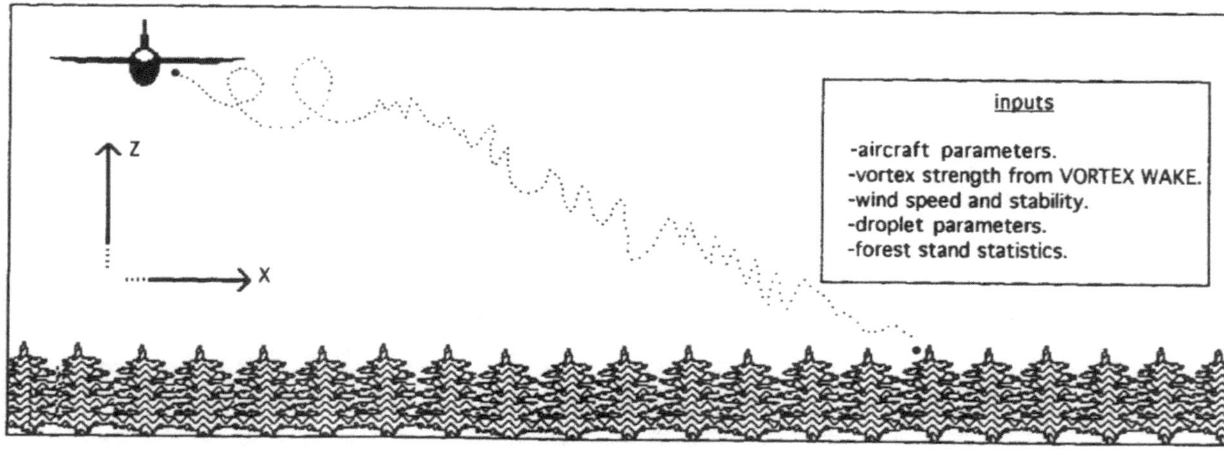

VORTEX CONTROLLED

inputs & calculations

-vortex wake strength→velocity field in vortices in crosswind.
-droplet settling and evaporation.
-droplet position from vortex velocity field and settling velocity of evaporating droplet.

TURBULENCE AND MEAN MOTION

inputs & calculations

-mean velocities in atmospheric boundary layer.
-vertical turbulent Lagrangian length and velocity scales.
-droplet settling and evaporation .
-droplet position from mean velocity, random turbulent displacement and settling.

FOLIAGE CAPTURE OR GROUND DEPOSITION

inputs & calculations

-foliage characteristic described by an x-z grid of cells.
-droplet reaches a cell, probability of capture established from foliage data, velocity, size of droplet,
-droplets not captured move to next cell, may drift further or sink to ground.

Figure 6.1 Outline of the PKBW modelling scheme

$$\Delta x = u(x,z)\,\Delta t$$
$$\Delta z = w(x,z)\,\Delta t \qquad (6.2.1)$$

When droplets are in the vortex, u and w (instantaneous velocities) values are obtained from the vortex model and the time step is fixed at 0.2 seconds. Once the droplets leave the vortex, trajectory calculations are continued with a one-dimensional turbulence model. Vertical turbulence fluctuations only are considered:

$$u = \mathrm{u}\,(z)$$
$$w = \mathrm{w}\,(z) + \mathrm{w}'(z) = w(z) \qquad (6.2.2)$$

Italicized u and w refer to instantaneous velocity, and the prime indicates turbulent fluctuations. The u(z) and w(z) indicate time-averaged values dependent upon height. Displacements are calculated as:

$$\Delta x = u(z)\,\Delta t$$
$$\Delta z = w(z)\,\Delta t \qquad (6.2.3)$$

Here, Δt is the Lagrangian time step, u(z) the mean horizontal component of the wind velocity (identical for both Eulerian and Lagrangian reference frames), and w(z) the instantaneous vertical component, which consists of two parts, Lagrangian turbulent velocity and droplet terminal settling velocity:

$$w(z) = w_L(t) + w_g \qquad (6.2.4)$$

The time step Δt is chosen to be one-tenth of the Lagrangian time scale τ_L, defined as:

$$\tau_L = \int_0^\infty R_L(z,\Delta t)\,d(\Delta t) \qquad (6.2.5)$$

where R_L is the Lagrangian autocorrelation function:

$$R_L(z,\Delta t) = \frac{\overline{u'(t,z)\,u'(t+\Delta t,z)}}{\overline{u'^2}} \qquad (6.2.6)$$

When $\Delta t \ll \tau_L$, the following approximation is valid:

$$R_L = \exp\left(\frac{-\Delta t}{\tau_L}\right) \qquad (6.2.7)$$

The Lagrangian turbulent velocity is constructed as a Markov chain process:

$$w_L(t) = w_L(t - \Delta t)\cdot \exp\left(\frac{-\Delta t}{\tau_L}\right) + \left(1 + \exp\left(\frac{-2\Delta t}{\tau_L}\right)\right)^{0.5} \cdot \sigma_w \cdot r(t) \qquad (6.2.8)$$

Here σ_w is the turbulent vertical velocity scale and r(t) is a random number from a Gaussian distribution with mean zero and standard deviation of unity. In addition to the terms in equation 6.2.4 , a correction term is needed to account for the fact that in situations with height-dependent velocity scale σ_w , the above prescriptions of vertical velocity will produce a net vertical flux of air. Wallace and Picot (1991) compare two methods for this correction, the Wilson bias velocity method and the LeClerc reflection probability method. PKBW permits a choice of the two methods.

6.1.3 VELOCITY AND TURBULENCE IN THE ATMOSPHERIC BOUNDARY LAYER

The following relations are used to carry out the computations expressed in equations 6.2.3 and 6.2.8:

For $z > z_T$ (mean tree height)

neutral atmosphere

$$u(z) = \frac{u*}{k}\cdot \ln\left(\frac{(z-z_d)}{z_o}\right)$$

$$\tau_L(z) = 0.5\frac{z}{\sigma_w(z)}$$

$$\sigma_w(z) = 1.25u*$$

stable atmosphere

$$u(z) = \frac{u*}{k}\cdot\left[\ln\frac{z}{z_o} + \frac{4.7}{L}\cdot(z-z_o)\right]$$

$$t_L(z) = 0.5\ \frac{z}{\sigma_w}(z)\cdot\left(1 + \frac{5.0z}{L}\right)^{-1}$$

$$\sigma_w(z) = 1.25u*$$

unstable atmosphere

$$u(z) = \int_{z_o}^{z} \frac{u*}{kz} \cdot \left(1 - \frac{16.0z}{L}\right)^{0.25} dz$$

$$\tau_L(z) = \frac{0.5z}{\sigma_w(z)} \cdot \left(1 - \frac{6.0z}{L}\right)^{0.25}$$

$$\sigma_w(z) = 1.25u* \cdot \left(1 - \frac{4.1z}{L}\right)^{0.33}$$

Symbols used above are defined in section 2.2. For z below z_T, the following relations are used:

$$u(z) = u(z_T)\exp\left[-\alpha\left(1 - \frac{z}{z_T}\right)\right] \qquad (6.2.10)$$

The value used for α is 0.25 for all stability classes. For values of τ_L and σ_w in the canopy region, the following are recommended:

neutral, for 1.2 z_T downwards:

$$\tau_L(z) = \tau_L(1.2\ z_T)$$

and for z_T downwards:

$$\sigma_w(z) = 1.25u* \cdot \left[0.2 + 0.8\exp\left(-4.6\left(\frac{z_T - z}{z_T}\right)^2\right)\right] \qquad (6.2.11)$$

stable, for 1.2 z_T downwards:

$$\tau_L(z) = \tau_L(1.2z_T)$$

$$\sigma_w(z) = 1.25u*$$

unstable, for 1.2z_T downwards:

$$\tau_L(z) = \tau_L(1.2\ z_T)$$

and for $2z_T$ downwards:

$$\sigma_w(z) = \frac{z}{2z_T} \cdot 1.25\ u_* \left[1 - \frac{4.1\, z_T}{L}\right]^{0.33}$$

Equations 6.2.11 were based on empirical fits to field data on σ_w and τ_E obtained by Mickle et al (1984) from the Dunphy field experiments. The relation suggested by Hall(1975) was used to obtain τ_L:

$$\tau_L \approx 4.5\, \tau_E \qquad (6.2.13)$$

6.1.4 CALCULATING DEPOSITION ON CANOPY AND GROUND

For each size class in the emitted droplet spectrum, a preselected number of droplets (1000 or more) is used to simulate the net trajectory pattern for the size class. Weighting factors are used to convert the preselected number to a value representative of the volume of the size class in the actual spray. An evaporation calculation following the methods described by equations 3.2.7 to 3.2.12 is done for each time step to provide a revised droplet size for each time step.

Droplet settling velocity w_g is computed from correlations given by Strauss(1975)(see Figure 2.4). Droplet locations are stored in cells of a horizontal-vertical grid system set up for the complete zone of interest for the calculation, from just upwind of the aircraft flight line to the limit of the downwind drift zone, and from ground to the mixing layer height. The tree canopy foliage distribution (see section **2.1**) is also stored in this grid. When a droplet enters the canopy zone, there is a probability that it will deposit on foliage. This probability is the product of a)the probability that the droplet trajectory will cause it to interact with a foliage element, and b) the probability that the interaction will result in deposition by inertial impaction or interception. The first of these is equivalent to the foliage cross-section per unit of spatial area through which the droplet is moving, and the second is the impaction efficiency for droplets on foliage shoots, including foliage orientation and foliage sheltering effects. This collection efficiency is a product of the May and Clifford (1967) ribbon collection efficiency (Figure 2.5) and a combined orientation and shelter factor given in Table 6.1 [Wallace et al (1995)]. Chapman (1993) has pointed out that for deciduous (broadleaf) foliage, the Stokes number for use in the May and Clifford data requires a leaf diameter equivalent to (leaf length+leaf width)/2 and an empirically obtained multiplier constant of 0.47.

The droplet deposition probability on foliage represents a number of droplets deposited when the spectrum weighting factor is used as a multiplier. Deposits are accumulated by grid location. Droplets not deposited on foliage are followed until they deposit on the ground or are carried beyond the downwind zone of interest by the wind (ground level is defined as 0.70 times roughness height to account for low-level ground cover plants). PKBW provides droplet deposit and spectrum for each downwind grid position, as well as the location, density, and droplet size spectrum for the drift cloud, all at 10 second intervals. The fraction of emitted spray material deposited on canopy and ground and off-target drift are also provided.

TABLE 6.1. Combined shelter and orientation factor from field calibrations

Dunphy, N.B. , site*			Sheephouse Brook, N.B. site	
Spruce	Jackpine	Fir	Maple	Raspberry
0.109	0.081	0.143	0.70**	0.83**

*the product of ave. f of Table 2.3 and $E_{\mu,i}/E_s$ of Table 2.4
** using leaf diameter as (length+width)/2 and multiplier of 0.47 in Stokes No.

6.1.5 INPUT DATA FOR A TYPICAL RUN

Table 6.2 below summarizes the input information provided to PKBW.

TABLE 6.2 List of input data for PKBW

1. Calculation run identifiers
2. Aircraft type and operating conditions for vortex behaviour, flying height
3. Stand description: location, roughness height, displacement height, mean height, base of foliage crown, foliage density and vertical distribution, effective height of ground cover (for up to three different stands in the spray zone)
4. Atmospheric conditions: temperature, pressure, Monin-Obukhov length (for stability class), wind velocity (friction velocity for at least one of the stands, allowing calculation for all stands), relative humidity, height of mixing layer
5. Atomizer identifiers, location on wing, number, flow rate, droplet spectrum
6. Formulation properties for evaporation calculations, % nonvolatile, tracer concentration
7. Calculation parameters: upwind and downwind grid limits, grid size, time limit for calculation

A graphical menu-driven input and output format is used with PKBW as well as printed versions of input and output files. Utility macros are used to set up block spraying simulations on spread sheets.

6.1.6 OUTPUT FOR THE EXAMPLE

Table 6.3 shows a typical "short report" printed out from a PKBW simulation of one of the Dunphy research field trials. A TBM Avenger spray aircraft fitted with 24 11010 Teejet atomizers was used in these trials. The liquid was an aqueous emulsion containing 17% non-volatiles and a tracer.

TABLE 6.3 PKBW simulation of the Dunphy #2(1986) TBM spray trial : short report

SIMULATION OF 4T86	CASE NO. 4	TBM AVENGER	NEUTRAL
DATE OF SIMULATION :		18 4 1996	
RELATED OUTPUT FILES:		4T286.CNT	4T286.SUS

AIRCRAFT AND ATOMIZER PARAMETERS:

AIRCRAFT TYPE =	TBM AVENGER	WINGSPAN=	16.5 m
AIRCRAFT SPEED =	77.0 M/S	AIRCRAFT HEIGHT =	45.0
SPREAD UNDER THE WING	6.0m	DISTANCE FROM FUSELAGE	1.5m
ATOMIZER TYPE =	TEEJET 11010	FLOW PER ATOMIZER=	3.9 l/min
NUMBER OF ATOMIZERS =	24	EMSN. OF A.I. OR TRACER=	429 mg/m

DROPLET SPECTRUM VOLUME FRACTION	DIAMETER (μm) (nominal)	(BEFORE EVAPORATION) (range)
.15	15.00	0-30
.12	40.00	30-50
.13	63.00	50-75
.14	88.00	75-100
.28	125.00	100-150
.12	175.00	150-200
.06	250.00	200-300

VORTEX TIME	VORTEX UPWIND CIRC.(m^2/s)	VORTEX DOWNWIND CIRC.
.00	65.23	65.23
1.00	63.29	57.03
2.00	25.63	18.81
3.00	20.23	8.99
4.00	15.46	3.84
5.00	9.82	1.41
6.00	6.03	.00
7.00	3.41	.00
8.00	1.05	.00
9.00	.00	.00

WEATHER PARAMETERS:

ATMOSPHERIC STABILITY CLASS= NEUTRAL

AIR TEMPERATURE = 15.20 Deg C ATM PRESSURE= 101.30 kPa

RELATIVE HUMIDITY= .78 MONIN-OBUKHOV LENGTH= -833.00m

STAND	FRICTION VELOCITY (m/s)	MIXING LAYER HT (m)
1.	.835	450.00

STAND PARAMETERS:

STAND NUMBER 1.

STAND TYPE= DUNPHY

ROUGHNESS HT.= 1.44m BASE OF THE FOL. CROWN= 7.50 m

MEAN TREE HT.= 15.00m DISPL'T HT. = 11.26m

NEEDLE OR LEAF D. = .20 E + 01 mm NEEDLE OR LEAF L. = .10 E + 02 mm

FOLIAGE DEN. = 2.94 SQUARE METERS OF FOLIAGE/SQUARE METERS OF GROUND

FORMULATION PARAMETERS:

CARRIER= WATER

MOL. WT. OF CARRIER = 18.00 g/mol DEN. OF CARRIER= .10E + 04 Kg/m^3

HEAT OF VAP.= .44 E + 08 J/mol DIFF. IN AIR= .23 E-04 m^2/sec

FRACTION NON-VOLATILE= .17 CONC. OF TRACER OF A.I.= 21200.0 mg/l

TOT. DEP. RESULTS (mgs OF TRACER OR A.I./SQ m GRND)

*SEG. #	**MDPT (m)	GROUND DEPOSITE	CANOPY DEPOSIT
1	-15	.0000	.0000
2	-5	.0000	.0000
3	5	.0000	.0000
4	15	.0000	.0000
5	25	.0000	.0000
6	65	.0000	.0000
7	45	.0000	.0000
8	55	.0000	.0000
9	35	.0187	.0327
10	75	.0080	.0211
11	85	.0418	.1271
12	95	.0568	.1185
13	105	.0574	.2871

* SEG # = DOWNWIND GRID POSITION LABEL
** MDPT = MIDPOINT OF GRID ELEMENT

14	115	.1282	.2276
15	125	.1056	.1737
16	135	.1044	.1890
17	145	.0604	.1631
18	155	.1209	.2795
19	165	.0949	.2708
20	175	.1601	.3072
21	185	.1705	.1988
22	195	.0562	.1815
23	205	.0697	.1148
24	215	.0367	.1494
25	225	.0636	.1556
26	235	.0937	.1519
27	245	.0954	.1960
28	255	.1051	.2256
29	265	.0542	.1918
30	275	.1044	.0992
31	285	.0340	.1801
32	295	.0420	.1197
33	305	.0221	.1183
34	315	.0447	.1578
35	325	.0871	.0996
36	335	.0614	.1001
37	345	.0438	.1462
38	355	.0369	.1218
39	365	.0540	.1195
40	375	.0346	.0785
41	385	.0167	.0697
42	395	.0242	.0908

INTEGRATED FLUX AT 400.0 METERS (mgs OF TRACER OR AI/m^2)

J*	ZGPSN(m)*	SUS(J)*	J	ZGPSN(m)	SUS(J)
1	1.00000	.26977	14	234.48000	.00000
2	18.96000	1.5989	15	252.44000	.00000
3	36.92000	4.6377	16	270.40000	.00000
4	54.88000	4.9239	17	288.36000	.00000
5	72.84000	4.2425	18	306.32000	.00000
6	90.80000	1.7094	19	324.28000	.00000
7	108.76000	1.3288	20	342.24000	.00000
8	126.72000	.88934	21	360.20000	.00000
9	144.68000	.09084	22	378.16000	.00000
10	162.64000	.04304	23	396.12000	.00000
11	180.60000	.00000	24	414.08000	.00000
12	198.56000	.00000	25	432.04000	.00000
13	216.52000	.00000	26	450.00000	.00000

*J =label for vertical grid, ZGPSN = vertical grid position, SUS(J) = suspended droplet tracer flux, mgs/m^2

1 RESULTS FOR DROPSIZE DIST. PER NEEDLE
% DROPSIZE (UM, DIAMETER AFTER EVAP.)

SEG#*	MDPT(m)*	#DRPS/ NEEDLE	8.3	22.1	34.8	48.6	69.0	96.6	137.9
1	-15.0	.00	.00	.00	.00	.00	.00	.00	.00
2	-5.0	.00	.00	.00	.00	.00	.00	.00	.00
3	5.0	.00	.00	.00	.00	.00	.00	.00	.00
4	15.0	.00	.00	.00	.00	.00	.00	.00	.00
5	25.0	.00	.00	.00	.00	.00	.00	.00	.00
6	35.0	.00	.00	.00	.00	.00	.00	.00	.00
7	45.0	.00	.00	.00	.00	.00	.00	.00	.00
8	55.0	.00	.00	.00	.00	.00	.00	.00	.00
9	65.0	.00	.00	.00	.00	.00	.00	.00	100.00
10	75.0	.00	.00	.00	.00	.00	55.62	11.67	32.71
11	85.0	.02	.00	.00	1.94	.00	69.87	16.70	11.49
12	95.0	.07	45.90	.00	33.52	9.53	2.14	5.35	3.57
13	105.0	.45	46.44	7.54	25.75	14.36	4.60	.72	.60
14	115.0	.61	68.35	14.74	8.52	2.83	4.66	.53	.38
15	125.0	.22	25.94	38.66	19.81	7.45	4.94	2.41	.79
16	135.0	.11	8.75	14.37	27.77	27.53	17.94	1.46	2.17
17	145.0	.14	7.34	36.85	28.72	9.87	12.23	4.33	.65
18	155.0	.28	5.96	50.66	18.80	9.54	11.44	3.15	.45
19	165.0	.34	32.16	34.46	6.27	13.28	11.90	1.64	.27
20	175.0	.99	65.51	17.81	9.75	2.15	3.91	.73	.14
21	185.0	.23	25.58	31.04	21.86	9.45	9.32	2.25	.50
22	195.0	.23	25.22	29.57	22.26	12.57	7.92	2.11	.35
23	205.0	.28	42.50	33.23	13.79	4.48	4.94	.93	.14
24	215.0	.35	51.35	25.71	15.54	2.79	2.62	1.76	.23
25	225.0	.30	18.29	54.71	19.03	2.04	3.65	2.03	.24
26	235.0	.51	69.90	14.78	6.06	5.13	3.35	.65	.13
27	245.0	1.86	89.97	4.96	2.46	.70	1.67	.22	.02
28	255.0	.87	74.99	12.21	6.51	.92	4.91	.38	.07
29	265.0	1.18	85.87	6.32	3.97	1.14	2.35	.22	.11
30	275.0	.31	63.52	23.67	6.69	2.27	2.62	.93	.30
31	285.0	.31	55.88	21.84	5.98	2.97	12.33	.90	.08
32	295.0	.43	64.35	26.52	2.75	1.55	3.98	.76	.09
33	305.0	.35	62.67	19.33	9.54	4.04	2.85	1.55	.02
34	315.0	.21	56.73	10.17	14.66	6.43	8.87	2.80	.33
35	325.0	.14	64.07	3.63	14.68	6.13	8.73	2.55	.21

* SEG = label for downwind horizontal position,
MDPT = midpoint of downwind segment

36	335.0	.27	72.34	5.75	11.82	3.57	5.83	.55	.13
37	345.0	.36	57.44	23.78	9.21	3.29	5.20	.85	.24
38	355.0	.37	55.44	11.06	22.82	6.95	3.45	.24	.04
39	365.0	.21	16.05	46.19	23.87	3.44	9.70	.63	.12
40	375.0	.30	64.62	15.67	13.04	2.66	3.78	.14	.10
41	385.0	.75	80.93	11.84	6.03	.11	.89	.18	.02
42	395.0	.20	40.49	23.58	19.54	9.82	6.19	.31	.06

FINAL RESULTS:

	GROUND	CANOPY	SUSPENSION (<400.00 M)	SUSPENSION (>400.0 M)	TOTAL
MGS/M:	23.11	52.68	1.80	352.03	429.6
%MASS:	5.37	12.26	.41	81.96	100.00

In the above table, "spread under the wing" means the length over which atomizers are mounted on each wing, and "distance from the fuselage" means the location of the first atomizer on the wing, measured from the fuselage center-line. Seven droplet spectrum volume fraction ranges are used to describe the atomizer output per run. When that is not sufficient, multiple runs covering different sections of the spectrum are done. The upwind and downwind circulation data refer to the upwind and downwind wings assuming flight is at right angles to the wind direction.

Weather parameter inputs require the friction velocity, the Monin-Obukhov length, and the mixing layer height, which can be obtained if temperature and wind velocity versus height information are known, using the equations in section 2.2.

The stand parameters shown are for the Dunphy test site which is treated as a uniform stand. Up to three stand types in succession downwind can be handled, and each can be composed of conifer and broadleaf foliage, provided the vertical foliage distribution is known or can be assumed with sufficient accuracy.

Deposit results for a single aircraft spray application are given for each downwind grid position which is identified by the location of its center-point. Both ground deposit and the total canopy deposit are given for each location. For a mixture of conifer and broadleaf foliage, each is provided with a separate deposit figure.

The integrated flux table refers to the wind-carried material which has drifted out of the zone of calculation, in this case 400 m downwind of the spray line, during the passage of the spray cloud. Values are given for each vertical grid step up to the mixing layer height, with the center of each position given by "ZGPSN" in meters above ground level.

The drop size distribution on the foliage is given for each downwind grid position with the assumption that evaporation is complete. The diameter values refer to the evaporated nominal sizes given as input e.g. the 15 μm droplet class in the emitted spray is equivalent to the evaporated droplet of 8.3 μm in this table.

The "FINAL RESULTS" table gives overall ground and canopy deposit in the calculation zone, -20 m to 400 m downwind of the spray line in this case expressed in terms of mg tracer or active ingredient per m of flight line averaged over the 400 m fetch. Suspension <400m means drift that has not reached the downwind limit of 400m in this calculation which was limited to 250 s of real time.

The simulation shown in Table 6.3 was performed on a PC equipped with a Pentium 90 micro-processor and required about 2 minutes of computation time. The calculation was based on 1000 droplets tracked for each spectrum size class. This is the lower limit which can be used with acceptable accuracy, which in this case produces a spread of about +/- 10% from the mean for repeated test calculations. This is greatly reduced when 1500 droplets are used, but is achieved at the expense of 50% more computational time and a greater random access memory requirement.

6.3 Model Validations and Comparisons: PKBW Model

Figures 6.2 and 6.3 show a comparison of Dunphy field trial data [van Vliet (1987)] and the PKBW prediction for test 2/86 (Grumman Avenger TBM aircraft) from Table 6.3 for canopy and ground deposits. The scatter in the computed results is due to the fact that only 1000 droplets were used for each size class in the simulation as mentioned above.

Agreement is very good for the canopy deposit, and not as good for the ground deposit. This is in part because in PKBW ground deposit is defined as droplets that settle below a level equivalent to 70% of the roughness height. This is done to avoid the need to characterize low-level vegetation such as ferns, etc. in terms of their collection efficiency and foliage area. The experimental procedure for ground deposit did not adequately reflect this ground structure in that two-foot square panels were used as platforms for the flat samplers used. Some of the low-level droplets were probably intercepted by nearby ground cover vegetation which was at a level above these panels.

Experimental error is evident in the Dunphy trials, since mass balance closures for all experiments ranged from 60% to 115% [Picot et al(1993)] with an average over the 20 experiments of 83% [van Vliet (1987)]. Mass balance closure values were found by adding measured canopy and ground deposits and downwind drift past 400 m and dividing this by the material emitted by the aircraft.

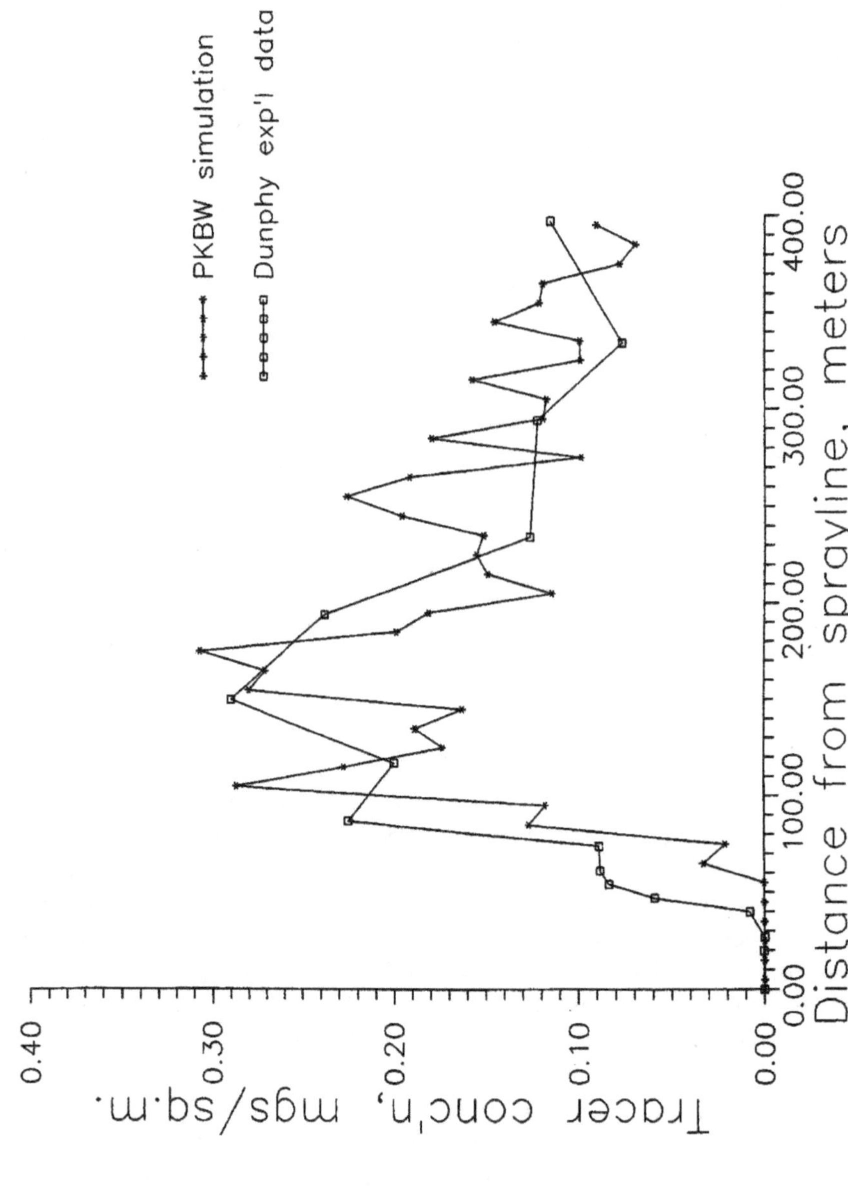

Figure 6.2 Comparison of experimental and PKBW simulation results for Dunphy trial run 2/86 (TBM) canopy deposits. Integrated deposits from sprayline to 400 m downwind are 14.3%(expt'l) and 12.3%(simulated).

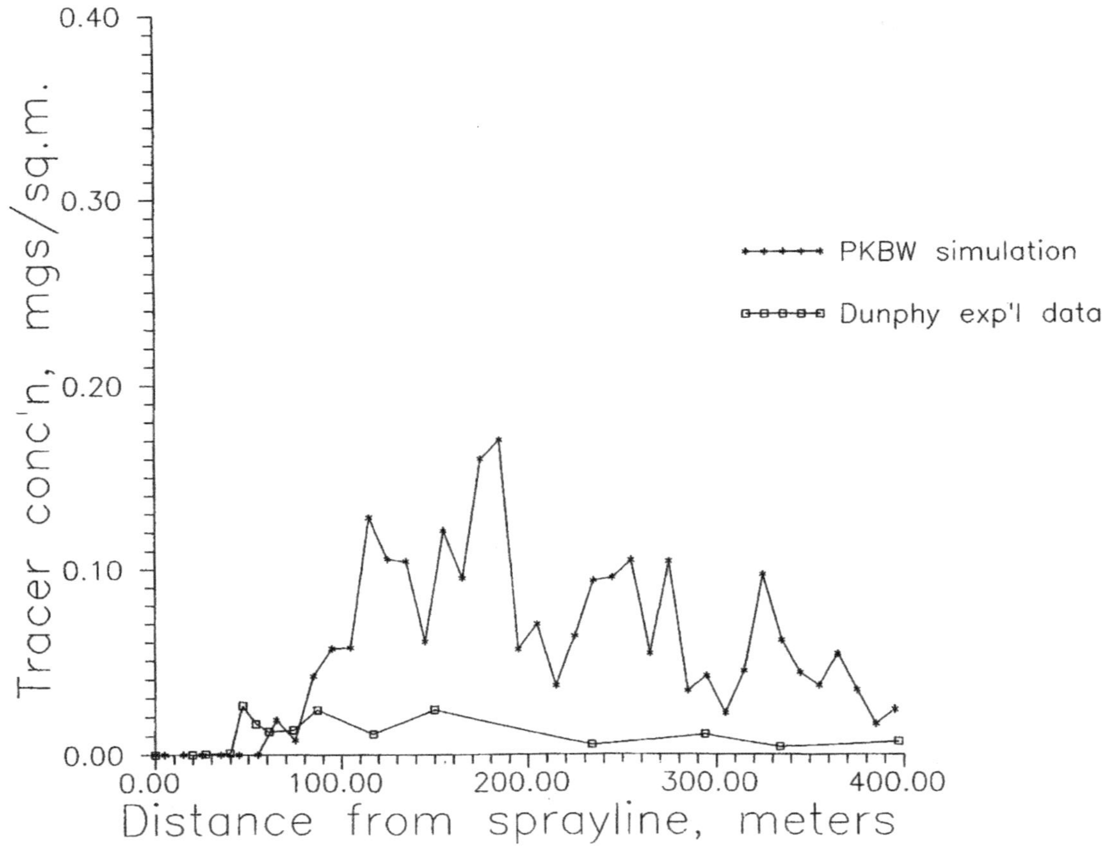

Figure 6.3 Comparison of experimental and PKBW simulation results for Dunphy trial run 2/86 (TBM) ground deposits. Integrated deposits from sprayline to 400 m downwind are 1.1% (exp'l) and 5.4% (simulated).

Environment Canada has commissioned a sensitivity analysis and validation of the AGDISP 6.1, FSCBG 4.3, and PKBW 2 spray drift and deposit models which was carried out by the New Brunswick Research and Productivity Council [Riley (1995)]. The field trial experiments used as bases for validation were four insecticide trials (Dunphy, N.B. 1986) and fourteen herbicide trials (Red Rock, N.B. 1989, and Sheephouse Brook, N.B. 1991). The insecticide scenarios for AGDISP and FSCBG were modelled without a canopy by using an apparent surface height of 12 m and a sample receptor height of 12 m, and reporting droplets reaching that height as total deposit. For PKBW, the Dunphy site foliage distribution characteristics available in the model were used. Canopy and ground deposits were added to give total deposit. The herbicide scenarios were modelled without a canopy in all cases. AGDISP and FSCBG models were provided with 26 droplet size categories, while PKBW used 7 categories only. Figures 6.4 to 6.7 show the results, normalized to a source strength of 1.0 gram of non-volatile tracer per meter of flight path. Typical input parameter listings are shown on Tables 6.4 and 6.5. The droplet spectra data used in the inputs are given in Appendix II.

Riley (1995) concludes in his study that all three models produced regression equations which predicted maximum deposit, the location of the maximum deposit, the integrated deposit, and the median integrated deposit location that differed from observed results by a factor of two or less. No model was consistently better or worse than the others and in this regard one model could not be recommended over the other two. Field trial values were assumed to be correct in these comparisons. The most important parameters affecting deposit profile were aircraft height, wind speed, and wind direction. Increasing the number of droplet size categories beyond 14 had no apparent systematic effect on the sensitivity of the three models.

6.4 Off-target Drift and Buffer Zone Estimation

6.4.1 SOME FIELD DATA

A field study of long-distance drift in forestry spraying was conducted in New Brunswick in 1979 by the National Research Council of Canada [Crabbe et al (1980)].Tree-top dosage of a simulated pesticide was measured at points 7.5, 24, and 45 km downwind of the aircraft spray line, over a homogeneous spruce-fir forest with a foliage density of approximately 3.0 m^2 per m^2 of ground. Linear relationship between the logarithm of dosage (droplet concentration integrated laterally across the plume and over the time of passage of the droplet cloud) and distance downwind was observed. It follows that droplet deposit should also

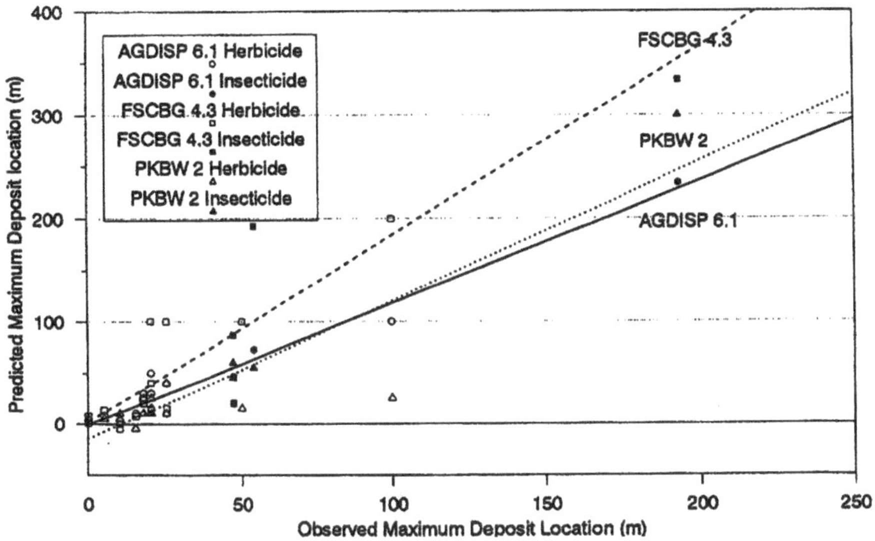

Figure 6.4 Predictions of three simulation models of the maximum total deposit for the conditions shown in Tables 6.4 and 6.5. Copied from Riley (1995) by the kind permission of Environment Canada.

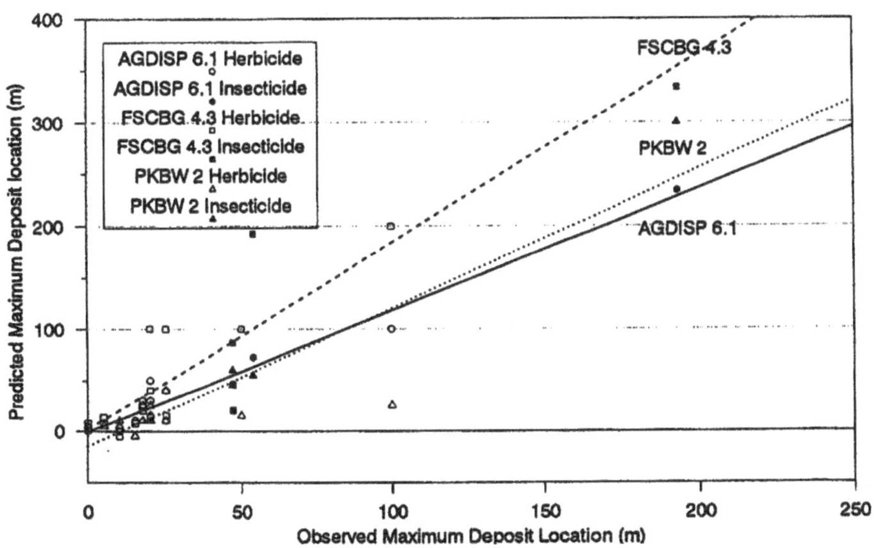

Figure 6.5 Predictions of three simulation models of the maximum deposit location on the canopy for the conditions shown in Tables 6.4 and 6.5. Copied from Riley (1995) by the kind permission of Environment Canada.

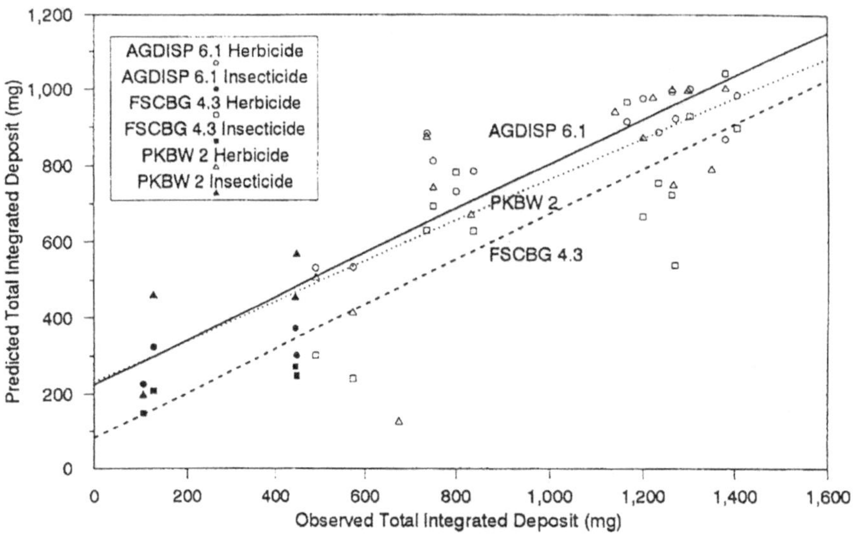

Figure 6.6 Predictions of three simulation models of the total integrated deposit for the conditions shown in Tables 6.4 and 6.5. Copied from Riley (1995) by the kind permission of Environment Canada.

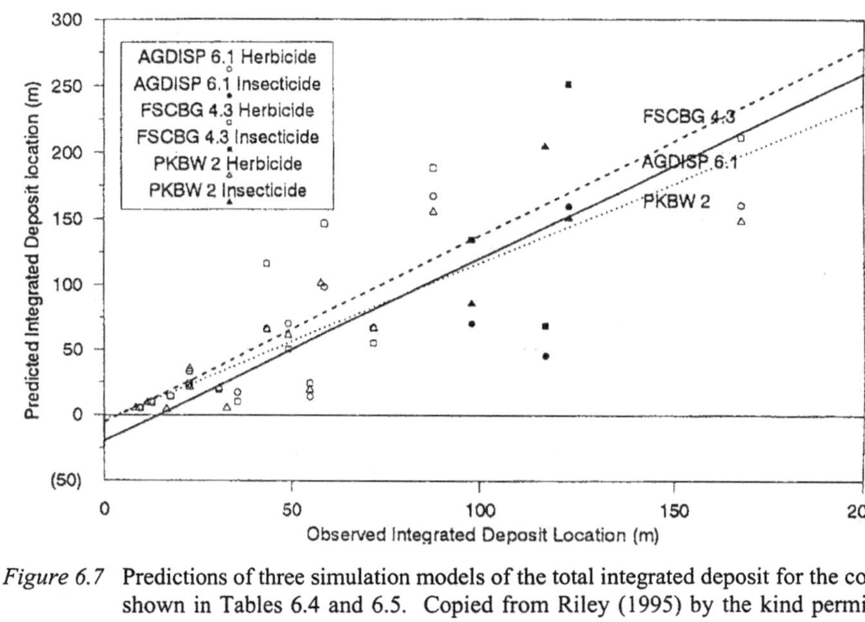

Figure 6.7 Predictions of three simulation models of the total integrated deposit for the conditions shown in Tables 6.4 and 6.5. Copied from Riley (1995) by the kind permission of Environment Canada.

TABLE 6.4 Typical herbicide application simulation input data list

Input Parameters for Database Reference # 1047/27_1991	
Maximum Time	600 sec
Aircraft Type	Cessna 188 Ag Truck
Aircraft Semispan	6.34 m
Nominal Release Height	29.0 m
Flight Speed	46.23 m/sec
Elliptically Loaded Wing	12322.5 N
Stability Class	neutral
Richardson Number	-0.02
Number of Crosswind Entries	5
Propeller Hub Height	-0.37 m
Axial Position	-3.63 m
Radius	1.1 m
Aircraft Drag Coefficient	0.1
Wing Planform Area	19.15 sq m
Efficiency	0.8
Shaft Rotation Rate	2850 rpm
Turbulence Fixed Value	0.0 sq m/sq sec
Number of Nozzles	28
Nominal Axial Position	0.0 m
Atomizer Horizontal Locations	-4.88, -4.57, -4.42, -3.96, -3.8, -3.5, -3.04, -2.6, -2.13, -1.82, -1.67, -1.2, -0.74, -0.28, 0.28, 0.74, 1.2, 1.67, 1.82, 2.13, 2.6, 3.04, 3.5, 3.8, 3.96, 4.42, 4.57, 4.88 m
Atomizer Vertical Locations	0.0m
Spray Mix Diluent	water
Specific Gravity	1.0
Concentration of Tracer	9730 mg/L
Spray System Flow Rate	26.76 gal/min
Volatile Fraction	0.94
Temperature	9.4°C
Temperature Depression	4.13 deg C
Air Pressure	101.3 kPa
Relative Humidity	52% @ 8.0 m
Friction Velocity	0.40 m/s
Mixing Layer Height	250 m
Monin Obukhov Length	-13 m
Circulation Decay Factor	0.56 m/sec
Stand Parameters	Sheephouse Brook
Spray Line Direction	270° to 90°
Droplet Spectrum	See table in Appendix B(ii)

Wind Speed	**Measurement Height (m)**	**Axial Velocity (m/s)**	**Crosswind Velocity (m/s)**	**Wind Speed (m/s)**	**Wind Direction (°)**
	2.17	-0.34	1.91	1.9	350
	4.14	-0.36	2.09	2.1	351
	8.04	-0.54	2.44	2.5	348
	16.40	-0.57	2.72	2.8	348
	18.80	-0.60	2.96	3.0	349

TABLE 6.5 Typical insecticide application simulation input data list

Input Parameters for Database Reference # 1049/07_1986	
Maximum Time	900 sec
Aircraft Type	Cessna 188 Ag Truck
Aircraft Semispan	6.34 m
Nominal Release Height	31.9 m
Flight Speed	49.0 m/sec
Rectangularly Loaded Wing	12322.5 N
Stability Class	neutral
Richardson Number	0.032
Crosswind Velocity	2.4 m/sec
Z	25.0 m
Zo	1.4 m
Direction	288 deg
Propeller Hub Height	-0.37 m
Axial Position	-3.36 m
Radius	1.1 m
Aircraft Drag Coefficient	0.1 m
Wing Planform Area	19.15 sq m
Efficiency	0.8
Shaft Rotation Rate	2850 rpm
Turbulence Fixed Value	0.4911 sq m/sq sec
Number of Nozzles	4
Nominal Axial Position	0.0 m
Atomizer Horizontal Locations	-3.51, -1.54, 1.54, 3.51 m
Atomizer Vertical Locations	0.0 m
Spray Mix Diluent	water
Specific Gravity	1.0
Concentration of Tracer	23663 mg/L
Spray System Flow Rate	7.6 gal/min
Temperature Depression	0.73 deg C
Volatile Fraction	0.832
Temperature	13.4°C @ 25 m
Relative Humidity	92% @ 25 m
Friction Velocity	0.36 m/s
Mixing Layer Height	250 m
Monin Obukhov Length	500
Circulation Decay Factor	0.5600 m/sec
Stand Parameters	Dunphy
Apparent Surface Height	12 m
Spray Line Direction	165° to 345°
Droplet Spectrum	See Table in Appendix B (ii)

follow an exponential decay at large distances downwind of the sprayline. Figure 6.8 below shows five graphs where the logarithm of canopy deposit is plotted against distance downwind, for data points downwind of the point where deposit

reaches about 50% of the peak deposit value (from the Dunphy studies).
The straight lines on the Figure 6.8 plots are curve fits for deposit as function of distance according to:

$$D = A.\ e^{-x/B} \tag{6.4.1}$$

where D is deposit in μg per g released per m of flight line. Regression analysis on the points shown on the graphs gives the following:

Form of 6.4.1 regressed:

$$\text{Log}(D) = \log A' - x/B' \tag{6.4.2}$$

The statistics are given below, using base 10 logarithms:

Run	Log(A′)	B′	Std err. of log(Dept) est.	R^2
6/86	2.71	1150	.253	.56
7/86	2.62	1220	.279	.45
8/86	2.98	752	.295	.69
9/86	2.57	637	.402	.62
10/86	2.83	833	.200	.80

For run 6/86 (Agtruck), deposits beyond the 1200 meter downwind can be adequately estimated by using 6.4.1 with the appropriate constants from the above table:

$$D = 10^{2.71}\ .\ 10^{-(x/1150)} \tag{6.4.3}$$

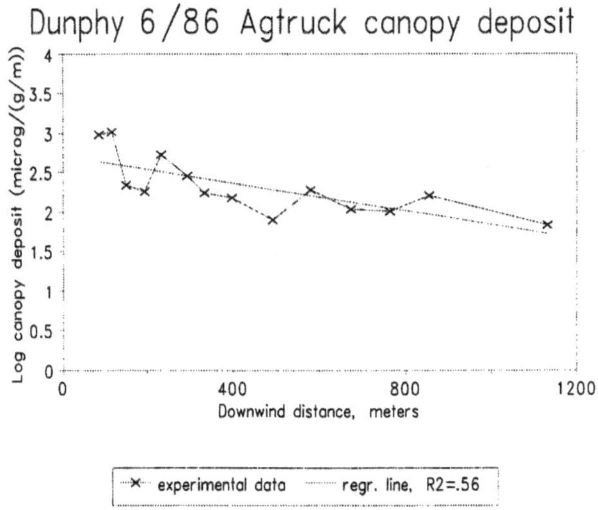

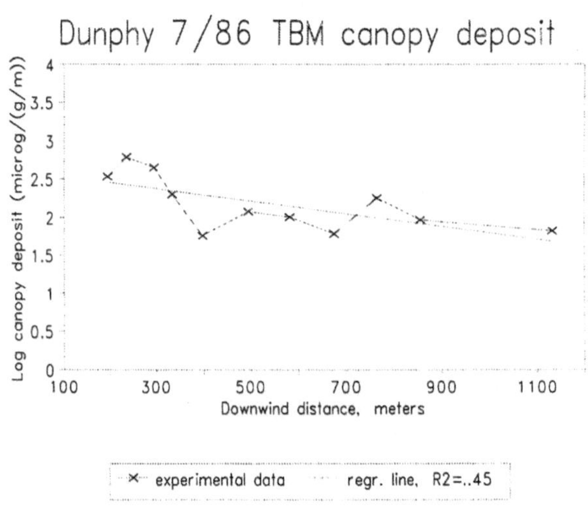

Figure 6.8 Assessments of exponential fits by regression analysis, Dunphy trial canopy deposit data

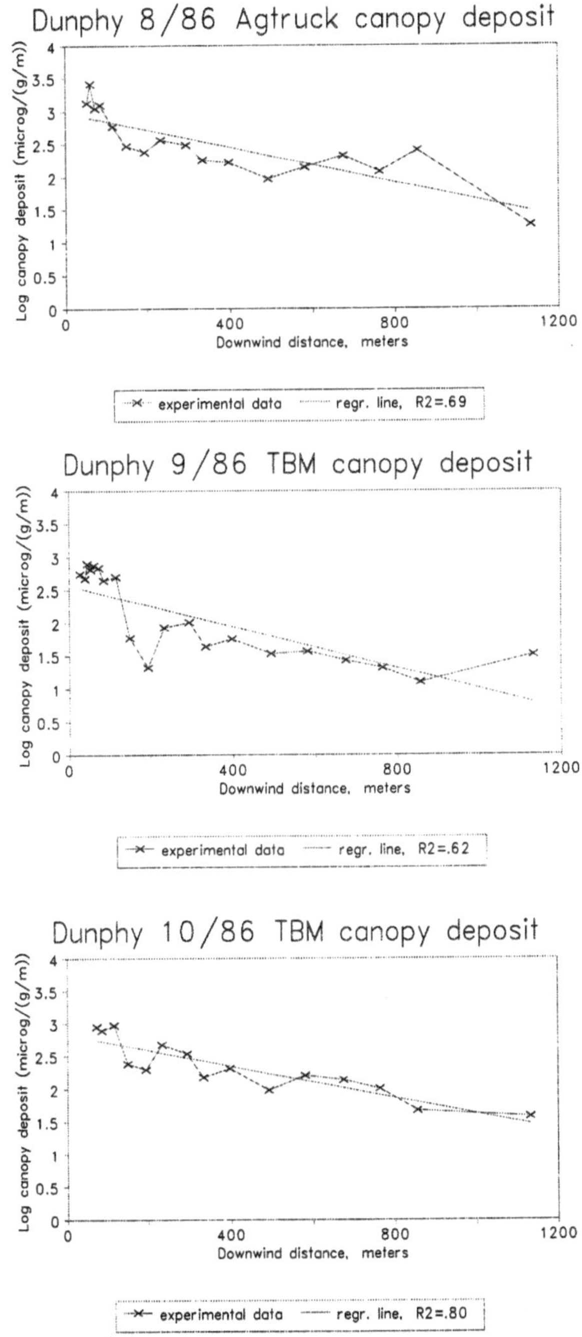

Figure 6.8 (cont) Assessments of exponential fits by regression analysis, Dunphy data

This equation can be used to show the deposit pattern within and beyond the measurement points, provided that weather and terrain parameters remain constant over the zone of extrapolation. The equation states that deposit tends to zero at infinite distance downwind, which is reasonable. The same form of equation will hold for ground deposit. A total deposit equation is obtained by adding canopy and ground deposit equations. That equation can be integrated with distance downwind to give an expression for cumulative downwind deposit as a function of distance downwind. Subtracting this from the aircraft emission rate provides an expression for downwind drift as a function of downwind distance from the aircraft application line.

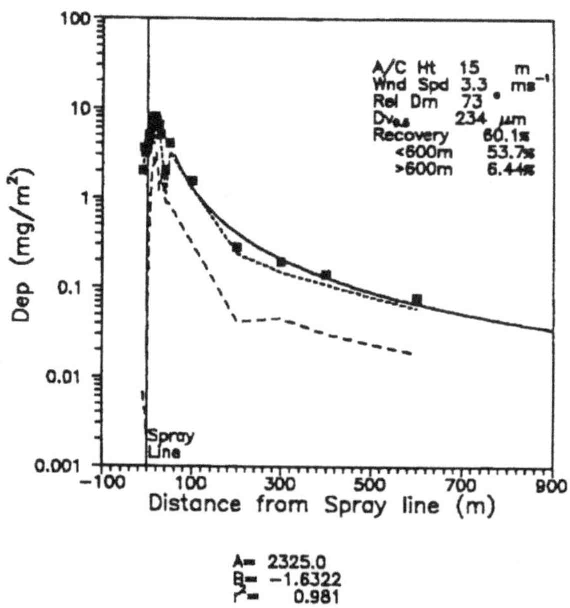

Figure 6.9 a. Total deposit data and curve fit from Mickle (1995), data ref. 1047/05/92. Copied by kind permission of the author. The solid line is the least squares fit to the data points for total deposit shown as squares. Other lines connect ground and foliage deposit data points respectively.

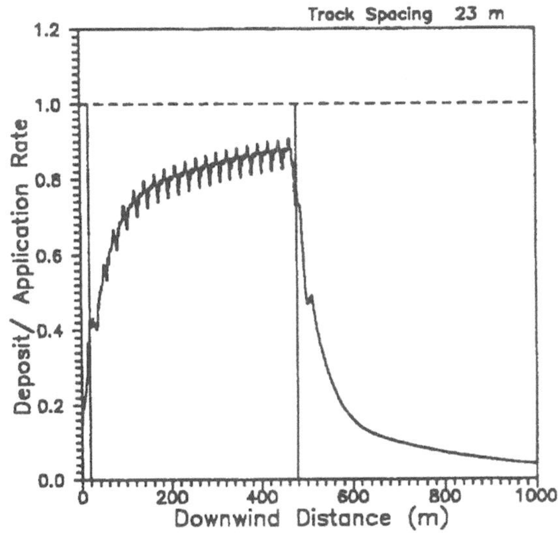

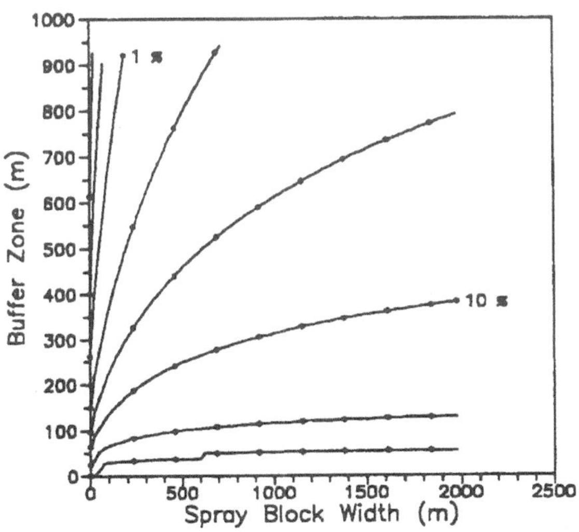

Figures 6.9b (top) and 6.9 c b)Normalized deposit vs downwind distance for a multiple-track application, and c)buffer zones required downwind of the last track to permit deposit to reach the values given by the curved lines showing 1%, 2.5%, 5%, 10%, 25%, and 50% of the application rate. Copied from Mickle (1995) by kind permission of the author.

Mickle (1995) has analyzed a large number of field trial data sets for the Canadian Interdepartmental Task Force on Pesticide Drift with the objective of developing a generic approach to establishing buffer zones. Total deposit was correlated with distance downwind using the relation:

$$D = A. (x)^{B} \tag{6.4.4}$$

Figure 6.9a shows results from this work for herbicide application tests done at Sevogle (Sheephouse Brook), N.B. in 1991. The aircraft used was a Cessna 188 Agtruck using D8-46 hollow cone nozzles. Other relevant data are shown on the graphic, along with the curve-fit constants and the correlation coefficient r^2. The relative direction, 73°, is the angle between aircraft flight vector and wind direction, and "recovery" refers to the fraction of emitted spray which was found on foliage, ground, and in the drift cloud. Figure 6.9 b shows a construction of the normalised deposit pattern for a multiple-swath application at a track spacing of 23 m under Figure 6.9a conditions. Figure 6.9c shows the results of swath-width calculations giving buffer zones required to allow off-target deposit to fall to prescribed percentages of the application rate.

6.4.2 USE OF PKBW FOR DRIFT ANALYSIS

Since field trial data are relatively sparse, and the quite accurate computer models mentioned above are available, specific questions relating to off-target drift are best addressed by the use of one of these models. This is illustrated below in Figures 6.10 a) and b) depicting PKBW results for drift of 15, 60, and 240 µm non-volatile droplets downwind of a spray line. The Dunphy stand characteristics(Table 6.3) and the test conditions of the Dunphy TBM 10-1986 [Picot et al (1993)] field trial were used (aircraft height 35 m, airspeed 77 m/s, u*=.51 m/s) except when indicated otherwise. The 15 micron results show the effect of aircraft height and stability class assuming the same u^* value. Except for the neutral condition, results are as expected, with the lower height unstable case giving the lowest drift at 2000 meters downwind. The neutral case result may be caused by an unusual interaction of velocity profile (which is affected by stability) and foliage profile. Using an average of the four depicted profiles and a regression of the results from 575 to 2000 meters, the following equation is generated:

$$\ln \{\text{Drift fraction}\} = -0.1486-0.000103 . x \tag{6.4.4}$$

Here, x is distance downwind of the spray line and ln is the natural logarithm. For a downwind location of 7500 meters, this gives a drift value of 40% of the original cloud of 15 micron, nonvolatile drops. For a typical TBM-TEEJETdroplet emission profile (Table 6.3), the zero to 15 micron size class after evaporation represents the zero to 30 micron size class on emission, approximately, and this accounts for the fraction 0.15 of the emitted mass. Since most of the >30micron size droplets have deposited before the 7500 meter point, the 40% drift of the 15 micron droplets relates to about 6% of the TBM-TEEJET emission (40% x 0.15). This estimate agrees very well with the Crabbe et al (1980) experimental result of 10% drift (average of one neutral and one slightly stable condition) under similar conditions.

The method outlined here with the PKBW model can be used to determine no-spray buffer areas when criteria for allowable off-target deposit and dosage from airborne pesticide are known. Given these criteria, PKBW can be used to assess spray application methods by comparing simulation results (based on proposed spray operating conditions) with the regulatory requirements. In addition to what has been shown here, PKBW can be used to plot the spray cloud droplet concentration as function of droplet size and horizontal and vertical position downstream.

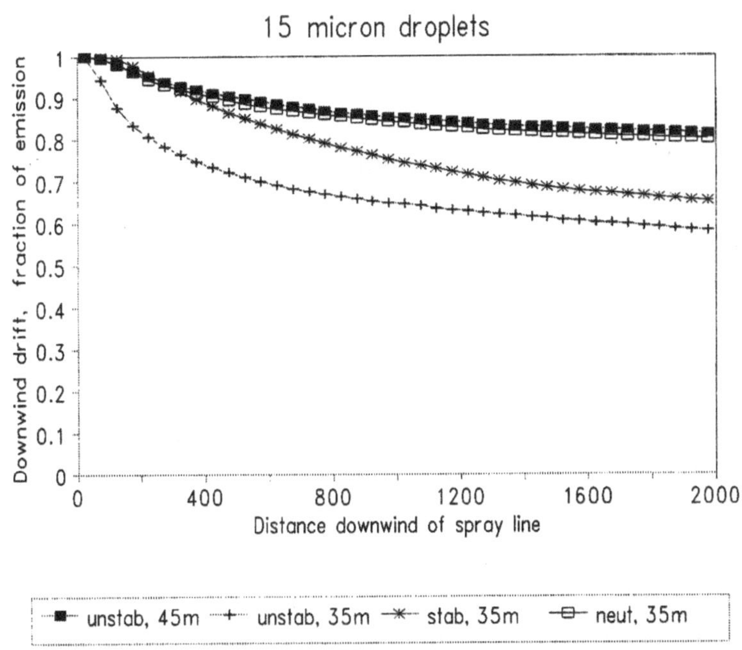

Figure 6.10 a) Fraction of hypothetical monodisperse spray cloud drifting down- wind, TBM aircraft, Dunphy forest, PKBW simulations. Monin-Obukhov lengths for stable and unstable conditions are 125 and -42 meters respectively. u^*=0.51 in all cases, 7000 droplets for each simulation.

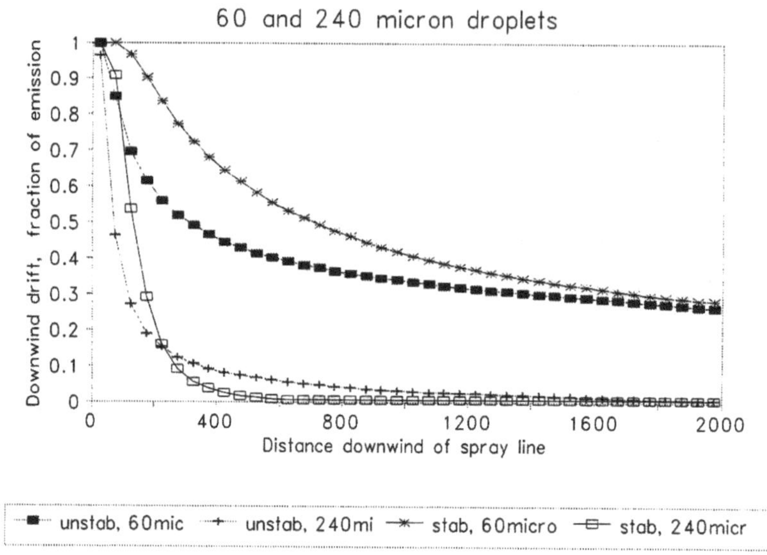

Figure 6.10 b) Fraction of two hypothetical monodisperse (one 60μm, other 240μm) spray cloud drifting down- wind, TBM aircraft, (35m height) Dunphy forest, PKBW simulations. Monin-Obukhov lengths for stable and unstable conditions are 125 and -42 meters respectively. u^*=0.51 in all cases, 7000 droplets for each simulation.

6.5 Simulation of Block Spraying, PKBW Model

Operational spraying of forested areas involves multi-track application of pesticide, often with two or three aircraft flying together at a predetermined spacing. A PKBW simulation assumes aircraft flight lines are at right angles to the wind direction with a uniform spacing between tracks. For the conditions given in Table 6.3 (except with an aircraft height of 35 meters), Figure 6.11 gives the result of an application with aircraft flight line spacing of 150 meters, for ten parallel tracks. The calculation is a simple addition of each deposit pattern, with each successive pattern offset from the previous one by a distance of 150 meters. The deposit pattern shows that 1300 meters are required before deposit becomes uniform, in this case, and deposit drops off sharply downwind of the last application line. The use of this type of simulation for the comparison of spray methodology alternatives is discussed in Chapter 8.

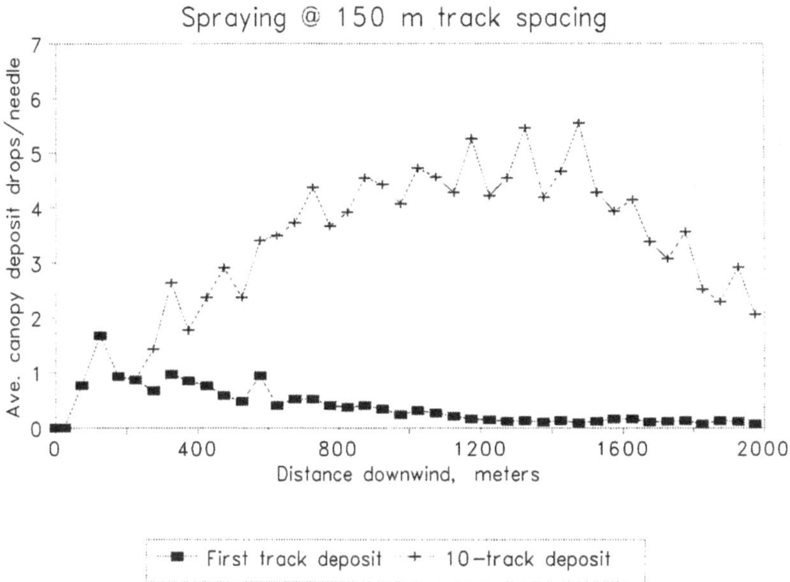

Figure 6.11 Spray block insecticide application by TBM aircraft (conditions as on Table 6.3 with 35 meter A/C height). The first application line is at the origin, followed by nine additional lines each spaced 150 meters downwind of the previous line. (Last line at 1350 meters downwind).

References

Atias, M., and D. Weihs, (1984) Motion of aircraft trailing vortices near the ground, AIAA Journal of Aircraft, **21**, pp 783-786.

Barry, J.W., M.E. Teske, J.A. Rafferty, and B.S. Grim, (1992): Predicting spray drift in complex terrain, ASAE paper #921085; ASAE, St. Joseph, MI.

Bilanin, A.J., M.E. Teske, J.W. Barry, and R.B. Ekblad, (1989): AGDISP: The aircraft spray dispersion model, code development and experimental validation, Trans. ASAE, **32**, pp 327-334.

Bilanin, A.J., M.E. Teske, J.W. Barry, and R. B. Ekblad, (1987): Comparisons of AGDISP code predictions with Program Wind deposition results, Proc. Symp. on the Aerial Applic'n of Pesticides in Forestry, NRC Canada, NRC #29197 (AFA-TN-18), Ottawa, pp 139-150.

Bilanin, A.J., and M.E. Teske, (1984): Numerical Studies of the Deposition of Material Released from Fixed and Rotary Wing Aircraft, NASA CR 3779.

Bilanin, A.J., M.E. Teske, and J.E. Hirsh, (1978): Neutral Atmospheric Effects on the Dissipation of Aircraft Vortex Wakes, AIAA Journal of Aircraft, **16,** pp 956-961.

Brashears, M.R., and J.N. Hallock, (1974): Aircraft wake vortex transport model, J. Aircraft, **11**, pp 265-272.

Chapman, T.J. (1993): Modelling of Herbicide Spray Dispersion and Deposition, M.Sc. Thesis, University of New Brunswick, Fredericton, N.B.

Crabbe, R.S., L. Elias, M. Krzymien, and S. Davie, (1980): New Brunswick Forestry Spray Operations: Field Study of the Effect of Atmospheric Stability on Long-Range Pesticide Drift, NRC Canada, Report LTR-UA-52, Ottawa.

Dumbauld, R.K., J.R. Bjorkland, and S.J. Satrelie, (1980): Computer Models for Predicting Aerial Dispersions and Deposition above and within Forest Canopies, USDA Forest Service Report 80-11.

Geary, H.V., Bowman, C.R., Rafferty, J.E., and Dumbauld, R.K. (1987): Application of the FSCBG version 2 aerial spray model in forestry, Proc. Symp. on the Aerial Applic'n of Pesticides in Forestry, NRC Canada #29197 (AFA-TN-18), Ottawa, pp 151-158.

Hall, C.D. (1974): The simulation of particle motion in the atmosphere by a random-walk method, Quart. J. R. Met. Soc., **101**, pp 235-244.

May, K.R. and R. Clifford, (1967): The impaction of aerosol particles on cylinders, spheres, ribbons, and discs, Ann. Occup. Hyg., **10**, pp 83-95.

Mickle, R.E. (1987): A review of models for ULV spraying scenarios, Proc. Symp. on the Aerial Applic'n of Pesticides in Forestry, NRC Canada, NRC #29197 (AFA-TN-18), Ottawa, pp 179-188.

Mickle, R.E., R.S. Crabbe, D.D. Kristmanson, J.J.C. Picot, J.D. Reid, B. Steeves, N.B.A. Trivett, and C.J. Wiesner, (1984): Assessment of the 1983 spray fate experiments at the Dunphy test site, Atmospheric Environmental Service Report # ARQB-84-010-L, Downsview, Ont.

Mickle, R.E., (1995) A generic approach to setting buffer zones in Canada, in Biorational Pest Control Agents, Formulation and Delivery, F.R. Hall and J.W. Barry eds., ACS Symposium Series, pp 27-40; Figures 8a,b,c taken from the original report to SERG of the same title, available at the Canadian Institute for Scientific and Technical Information, Ottawa.

Picot, J.J.C., D.D. Kristmanson, R.E. Mickle, R.B.B. Dickison, C.M. Riley, and C.J. Wiesner, (1993): Measurements of folial and ground deposits in forestry aerial spraying, Trans. ASAE, **36,** pp 1013-1024.

Picot, J.J.C., D.J. Wallace, and D.D. Kristmanson, (1987): The PKBW model for prediction of aerial spraying deposition and drift, Proc. Symp. on the Aerial Applic'n of Pesticides in Forestry, NRC Canada, #29197 (AFA-TN-18), Ottawa, pp 161-170.

Picot, J.J.C., D.D. Kristmanson, and N. Basak-Brown, (1986): Canopy deposit and off-target drift in forestry aerial spraying: The effects of operational parameters, Trans. ASAE, **29**, pp 90-96.

Riley, C.M. (1995): A Sensitivity Analysis and Validation of the AGDISP 6.1, FSCBG 4.3, and PKBW 2 Spray Drift and Deposit Models, Report to Atmospheric Environment Service Canada, NB Research and Productivity Council Report #B/95/0008, Fredericton, N.B.

Strauss, W. (1975): Industrial Gas Cleaning, 2nd. ed., Pergamon Press, Oxford, England.

Teske, M.E., and J.W. Barry, (1993): Parametric sensitivity in aerial application, Trans. ASAE, **36,** pp 27-33.

Teske, M.E., K.P. Bentson, R.E. Sandquist, J.W. Barry, and R.E. Ekblad, (1991): Comparison of FSCBG model predictions with heather seed orchard deposition data, J. Appl. Met. **30,** pp1366-1375.

Teske, M.E. (1986): User Manual Extension for the Computer Code AGDISP MOD 4.0, USDA Forest Service, Contract 85-003.

Teske, M.E. (1990): AGDISP: Evaluation and validation training: User group code enhancements, USDA Forest Service, FPM 90-1.

van Vliet, S.J. (1987): Aerial Spraying of Forests: Measurements of Ground and Canopy Deposits, B.Sc. Thesis, Dept. Of Chemical Engineering, UNB, Fredericton, N.B.

Wallace, D.J., J.J.C. Picot, and T.J. Chapman, (1995): A numerical model for forestry spraying, Agric. and Forest Met., **76,** pp 19-40.

Wallace, D.J., and J.J.C. Picot, (1991): A comparison of two Lagrangian dispersion models applied to forest canopies, Agric. and Forest Met.,**53**, 337-346.

Wallace, D.J. (1988): Aircraft Vortex Effects in Forestry Aerial Spraying, Ph. D. Thesis, University of New Brunswick, Fredericton, N.B.

Weihs, D., and M. Atias, (1987): Calculation of aerosol distribution in aerial spraying of forested terrain, Proc. Symp. on the Aerial Application of Pesticides in Forestry, National Research Council Canada, NRC #29197 (AFA-TN-18), Ottawa, pp 171-177.

CHAPTER 7

SPRAYING IN COMPLEX TERRAIN

7.1 Introduction

When terrain to be sprayed is sufficiently inclined to the horizontal and non-uniform in structure, it can create its own micrometeorology, depending on the insolation. The USDA Forest Service has studied these flows and their influences on aerial spraying over a number of years, and we summarize below the results of their recent work.

7.2 Air Flows in Well-defined Mountain Valleys

When undisturbed by external influences such as cloudiness or high upper-level winds, many valleys undergo a regular diurnal evolution in wind pattern [Ekblad et al (1989), Whiteman(1990)]. The second author describes wind structure development for a deep valley on the western slope of the Rocky Mountains (U.S.). Figure 7.1 and 7.2 describe the structure of the winds which develop over a 3.5 to 5 hour period after sunrise from a well-developed temperature inversion caused by overnight radiative cooling. Direct sunlight causes the development of an up-valley convective boundary layer over the valley floor. The same is true for the valley sidewall. Up-slope winds can reach up to 3 m/s. A stable core remains above the valley floor convective layer and between the sidewall layers, with a down-valley flow persisting, but with a continuous eroding of its boundaries by the action of the convective flows. This process gradually leads to destruction of the stable core, depicted on Figure 7.2. Cross-valley advection can occur when one of the valley walls is more strongly heated by sunlight than the other.

7.3 Implications with Respect to Pesticide Spraying

We quote the following conclusions from Whiteman(op. cit.):

1) Spray released before sunrise or over shaded slopes will be released into shallow stable boundary layers over the slopes. Drainage flows in these boundary layers will be downslope. The overlying down-valley flows typically superimpose a down-valley wind component on these layers,

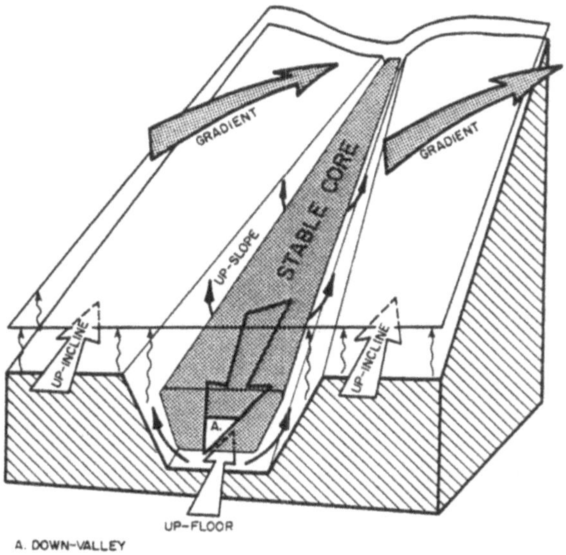

Figure 7.1 Typical mid-morning wind structure over and within a deep valley on the western slope of the Rockies, showing the five interrelated wind systems identified in field studies [from Whiteman (1990)]. The dark region represents the stable core moving down-slope, and the gradient winds are the winds in the free atmosphere above the valley. Copied with the kind permission of the USDA Forest Service.

so that any drift of the spray will move down-slope and down-valley. Pre-sunrise spraying may adversely affect population centers or sensitive areas that are downslope or down-valley during this time period. When plume impacts occur, concentrations may be quite high because of the poor diffusion environment along the transport path in the remnants of the strong nighttime inversion.

2) Initially, shallow convective boundary layers form over sunlit slopes. Shortly after they are sunlit (typically 10 or 15 minutes), an upslope wind will form. Drift of a non-deposited plume will then be upslope. If the plume escapes from the boundary layer (or if the spraying occurs too high above the slope), it will be transported down the valley on the remnants of the stable core.

3) Convective boundary layers develop rapidly over the ridgetops and upper

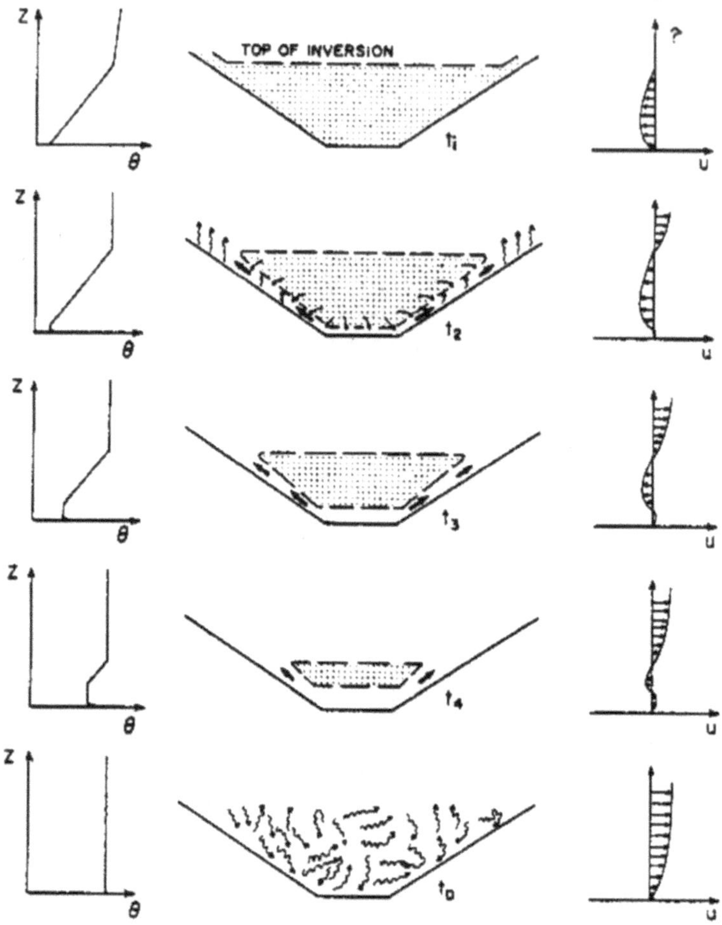

Figure 7.2 Illustration of the Hypothesis of Inversion Destruction. In the center of the diagram cross sections of a valley are shown at times t_i, t_2, t_3, t_4, and t_D. On the left are corresponding potential temperature profiles as taken from the valley center. On the right are corresponding up-valley wind components (u) as a function of height. At sunrise, t_i, an inversion is present in the valley At t_2, a time after sunlight has illuminated the valley floor and slopes, a growing convective boundary layer (CBL) is present over the valley surfaces. Mass and heat are entrained into the CBLs from the stable core above and carried up the sidewalls in the upslope flows. This results in a sinking of the stable core and growth of the CBLs (t_3 and t_4) until the inversion is broken (t_D) and a turbulent, well-mixed neutral atmosphere prevails throughout the valley depth. Down-valley winds continue to blow in the stable core during the inversion breakup period. Winds in the CBL below and in the region above the stable core often blow up-valley during this same period. [from Whiteman(1990)]. Copied by kind permission of the USDA Forest Service.

slopes after they become sunlit, because they are not capped by the strongly stable nocturnal inversion. Especially rapid convective boundary layer development may occur over the upper slopes that face the morning sun. When upper winds are moderate or strong, growth of the convective boundary layers will bring the stronger winds down onto the upper slopes. The upper slopes would then become more and more turbulent, with the normal thermal convection processes being supplemented by transport of horizontal momentum aloft. The rate of convective boundary layer development over the upper slopes will strongly affect the spraying environment there.

4) Once temperature inversion destruction begins, the top of the temperature inversion will sink into the valley, exposing more of the upper slopes to intense convective boundary layer development. Winds within the stable core during this time will continue to blow down the valley, but the speed will typically be decreasing until, just before the inversion is destroyed (3.5 to 5 hours after astronomical sunrise on a clear day), the winds will switch to up-valley throughout the valley's volume.

5) Given the above processes, rates of convective boundary layer development depend strongly on the local amount of insolation (which varies from place to place within the valley), the surface energy budget, and the overlying temperature structure. The ridgetops and upper slopes should be sprayed first, because the dispersion environment there rapidly becomes too turbulent for continued spraying. The spraying environment on the lower slopes, in contrast, is suitable for spraying for a much longer time. Spraying conditions can be quite different on opposing slopes during the morning transition period, especially in north-south oriented valleys where the opposing slopes have quite different heating functions. The east-facing slope will receive solar input earlier, the convective boundary layer will develop earlier and more rapidly, and the differential heating of the two slopes may result in a cross-valley flow toward the heated slope.

6) After the temperature inversion is destroyed, the valley becomes well coupled with the winds aloft. If these winds are weak, a thermally driven up-valley circulation will prevail within the valley. The absence of a strong capping inversion above the surfaces will result in a deep and turbulent convective boundary layer over all the valley's surfaces. If the upper winds are strong, they may superimpose their own wind direction on the valley, and winds may be correspondingly stronger and more turbulent, making conditions unsuitable for aerial spraying operations.

Below, we describe a dispersion model developed for the prediction of pesticide transport in a valley.

7.4 Predicting Pesticide Dispersion and Deposition in Valleys

Continuing earlier efforts to understand and model pesticide transport in complex terrain [Grim et al. (1992), Barry et al. (1992)], the USDA Forest Service initiated an agreement with the U.S. Department of Energy to develop a model to predict pesticide dispersion and deposition in valleys [Allwine et al. (1995)]. The main features and capabilities of this model, VALDRIFT, are described below.

VALDRIFT treats the time-dependent transport, diffusion, and deposition of an inert substance released from elevated or ground level sources into a valley atmosphere. Sources can be gases or aerosols having negligible settling velocities. The model addresses the drift of the fine droplet component of pesticide sprays after application and throughout the day after an early morning application. In addition to source characteristics and deposition, the physical processes treated are:

1. Along-valley flows
2. Turbulent diffusion
3. Nocturnal temperature inversion breakup
4. Convective boundary layer growth
5. Inversion descent
6. Subsidence

The model solves a one-dimensional (along-valley) species conservation equation for each of a number of "flow tubes" aligned along, and conforming to the shape the valley (Figure 7.3). Interactions can occur between flow tubes. The equation is as follows:

$$\frac{\partial(AC)}{\partial t} + \frac{\partial(VC)}{\partial S} - \Gamma_{cs} - \Gamma_{cp} - \Gamma_{cc} - \Gamma_{ca} - \Gamma_{cd} = 0 \qquad (7.4.1)$$

where t = time (s)

S = the along-valley coordinate (m)

C = the concentration along the flow tube (g/m^3)

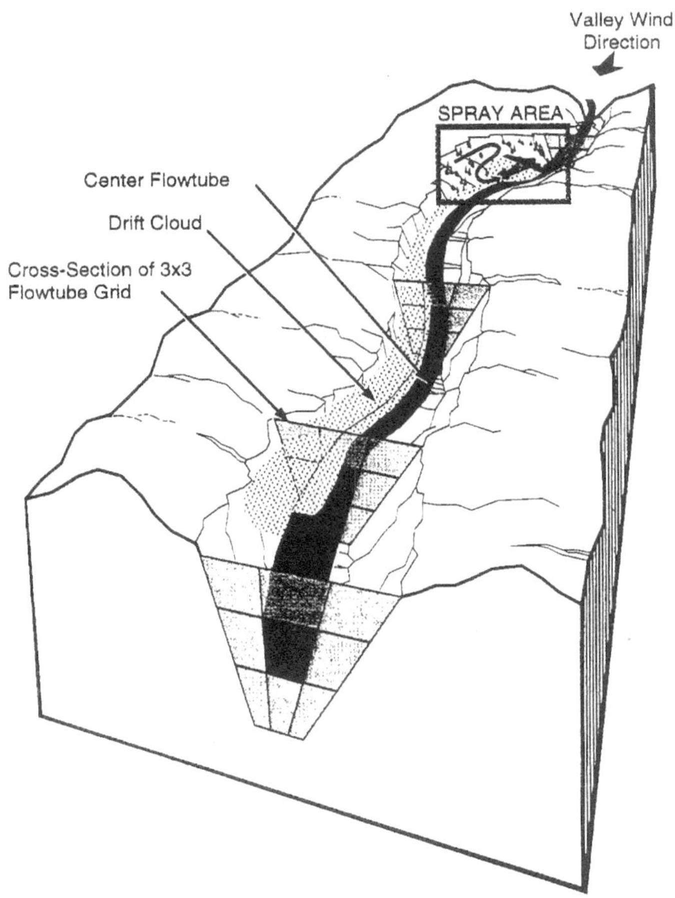

Figure 7.3 Schematic of flow tube arrangement for a VALDRIFT simulation [from Allwine et al. (1995)]. Copied by kind permission of the USDA Forest Service.

A = the cross-sectional area of the flow tube (m^2), a function only of S

V = the along-valley air volume flow rate through the flow tube (m^3/s)

Γ_{cs} = the emission source/sink term for the flow tube (g/s-m)

Γ_{cp} = the deposition source/sink term for the flow tube (g/s-m)

Γ_{cc} = the chemical transformations source/sink term for the flow tube (g/s-m)

Γ_{ca} = the lateral and vertical advection source/sink term for the flow tube (g/s-m)

Γ_{cd} = the turbulent diffusion source/sink term for the flow tube (g/s-m)

Each of the terms has dimensions of mass per unit time per unit length of flow tube. The first and second terms represent rate of accumulation and net rate of advection, respectively. In addition to Γ_{cs}, the Γ_{cp}, and Γ_{cd} terms have been implemented, as follows:

Γ_{cp} : The current version of this term is that for dry deposition, given by:

$$\Gamma_{cp} = V_d \, C \, \Delta A \tag{7.4.2}$$

where V_d = dry deposition velocity, $(u^*_d)^2 / U$
u^*_d = surface friction velocity for deposition
U = along-valley wind speed
ΔA = ground surface area of the cell divided by along-valley step size ΔS

Γ_{cd} : Turbulent diffusion in the lateral and vertical directions, viz.:

$$\Gamma_{cd} = \Gamma_{cdY} + \Gamma_{cdZ}$$

where $\Gamma_{cdY} = \Delta Z_i \; K_{Y,i} \left\{ \frac{\partial C}{\partial Y}\Big|_{\text{Face } j} - \frac{\partial C}{\partial Y}\Big|_{\text{Face } j-1} \right\}$;and

$$\Gamma_{cdZ} = \Delta Y_i \; K_{Z,i} \left\{ \frac{\partial C}{\partial Z}\Big|_{\text{Face } i} - \frac{\partial C}{\partial Z}\Big|_{\text{Face } i-1} \right\} \tag{7.4.3}$$

The K_Y and K_Z terms are lateral and vertical turbulent diffusivities which are functions of position, as are the step sizes ΔY and ΔZ. Published correlations are available for the diffusivities in stable, neutral, and unstable conditions [Businger et al. (1971), Panofsky et al. (1984), Bian et al. (1992)].

Equation 7.4.1 is solved by an explicit finite difference method but, first, requires a solution of the air mass conservation equation for each flow tube. This is done by treating the along-valley flow explicitly and treating nonrecirculating lateral flows (subsidence flows, regional flow intrusions, and tributary flows) as source/sink terms. The along-valley flow is described semi-empirically [Clements et al.(1989)] as a drainage flow (the stable core flow) dependent on position and height in the valley. A third calculation is required to specify the growth of the convective boundary layer and the descent of the stable core of the down-valley flow. This is based on the Whiteman and McKee (1982) thermodynamic model which describes the effect of solar heat flux over the valley surfaces. The model is applicable to well-defined valleys with steep sidewalls (10° to 90°). (The S,Y,Z grid system is generated with the S-axis following the valley floor centerline in the down-valley direction, the Y-axis is horizontal and locally perpendicular to the S-axis, and the Z-axis is oriented vertically).

VALDRIFT is a Fortran program. Memory requirements are about 4Mb for a 200x31x31 computational grid. Inputs required are terrain structure, meteorological parameters, and emission source information. Outputs include a summary of run parameters and the computed concentration and deposition information over the computational domain, as a function of time.

The completion of this program is considered to be an important step in improving predictions of pesticide drift for complex terrain spraying situations. The eventual objective is to incorporate VALDRIFT into the USDA Forest Service FSCBG modelling framework.

7.5 Comment

Currently, the eventual destination of off-target drift from aerial spraying operations is of prime concern. "No-spray" or "buffer" zones are usually established by regulatory agencies to protect sensitive areas. Complex terrain can produce complex drift patterns as shown above. Experienced applicators are aware of this situation and have adopted techniques for identifying potential problems (e.g. the use of smoke generators to identify the onset of seabreeze penetration over hilly terrain near a shoreline, often encountered in pesticide spraying in New Zealand).

References

Allwine, K.J., X. Bian, and C.D. Whiteman (1995): VALDRIFT 1.0- A Valley Atmospheric Dispersion Model with Deposition, PNL-10586, UC-402, a report to the USDA Forest Service, Missoula Technology and Development Center, from Pacific Northwest Laboratory, U.S. Department of Energy.

Barry, J.W., M.E. Teske, J.A. Rafferty, B.S. Grim, and P.J. Skyler (1992): Predicting Spray Drift in Complex Terrain, paper #921085, 1992 Int'l summer meeting, A.S.A.E., Charlotte, N.C.

Bian,X., E.S. Takle, and S. Zhong (1992): A theoretical study of the flux-profile relationships: Extension of the KEYPS equation to the tower layer, Preprints, 10th Symposium on Turbulence and Diffusion, 29 Sept.-2 Oct. 1992, Portland, Oregon, pp184-187, (Amer. Meteor. Soc., Boston, pub.).

Businger, J.A., J.C. Wyngaard, Y. Izumi, and E.F. Bradley (1971): Flux profile relations in the atmospheric surface layer, J. Atmos. Sci. **28**, pp 181-189.

Clements, W.E., J.A. Archuleta, and P.H. Gudiksen (1989): Experimental design of the 1984 ASCOT field study, J. Appl. Meteor. **28,** pp 405-413.

Ekblad, R.B., R.B. Hauser, and J.W. Barry (1989): Program Wind: Event Modelling for Complex Terrain (EMCOT), USDA Forest Service, Technology & Development Center, Missoula, MT, #MTDC 89-8.

Grim, B., J. Rafferty, G. Sutton, and T. Clarke (1992): Deposition of Bacillus Thuringiensis into Gambel Oak Canopies, USDA Forest Service, Forest Pest Management, Davis, CA, #FPM 92-9.

Panofsky, H.A., and J.A. Dutton (1984): Atmospheric Turbulence. Models and Methods for Engineering Applications., John Wiley and Sons, New York, 397 pp.

Whiteman, C.D. (1990): Effects of Valley Meteorology on Forest Pesticide Spraying, Pacific Northwest Laboratory (U.S. Dept. Of Energy) report PNL-7332; (included as Appendix A of report PNL-10586 (see above)), Richland, Washington.

Whiteman, C.D., and T.B. McKee (1982): Breakup of temperature inversions in deep mountain valleys: Pt II. Thermodynamic model, J. Appl. Meteor. **21**, pp 290-302.

CHAPTER 8

SPRAYING INNOVATIONS

8.1 Introduction

In the first three sections below we discuss the results of innovative developments in 1) droplet size optimization, 2) design of field trials for dose efficacy measurements, and 3) aircraft navigation and data logging. In the fourth section, we discuss possible improvements in application technology.

8.2 Drop Size Optimization

8.2.1 THE CONCEPT

Following up on the observations of Himel and Moore (1969) and others, Irving et al. (1990) described the concept of optimal droplet size for spraying spruce and fir forests to protect against defoliation by the spruce budworm. This concept arises from observations that 1) conventional spray application volumes are such that very few droplets greater than 60 microns are to be found on foliage even though a substantial fraction of the mass of the emitted spray is in the form of larger droplets; 2) atomizer characterization measurements show that droplet numbers increase with decreasing size; 3) capture efficiencies in the field are such that droplet numbers on foliage decrease rapidly with size below about 20 microns; and 4) for droplet sizes of approximately 20 microns and larger, modern insecticide potencies are such that defoliation by larvae is strongly inhibited at levels of one or more droplets per needle [Kettela and Steel (1990); van Frankenhuyzen (1995)]. As a consequence, an atomizer should be operated to maximize the cumulative volume fraction of droplets in the 15 to 55 μm range. This range is somewhat arbitrary, but encompasses the range of most frequent sizes deposited on needles, with recognition of the fact that for potent insecticides, even 15 μm droplets have a useful level of efficacy.

This optimized spraying has been called UULV spraying (ultimate ultra low volume?). Simply described, it is the production and distribution of only those droplets which have sufficient potency and are present in sufficient numbers to be efficacious (to the extent possible). Because droplet volume is proportional to the cube of droplet diameter, the use of large droplets can be hugely wasteful of pesticide if smaller droplets are effective. Assuming that larval foraging activity is a constant, for the same mass of insecticide, the probability of interaction between larvae and droplets will be eight times greater for 25 micron droplets than for 50 micron droplets (supposing uniform size droplets).

Irving et al. (op. cit.) go on to describe the performance of a new atomizer, the SERG/ONAIR system [Picot (1990); US Patent #4,948,050; Can. Patent #2,018,216-4; see Figure 4.5d], and compare it against conventional atomizers in terms of 1) number of droplets per litre generated in the 15 to 55 micron range, and 2) ratios of optimal size droplet mass to undesirable size droplet mass. This is shown on Table 8.1. The results of wind-tunnel testing (complete plume survey) show that the SERG/ONAIR atomizer operating at or near 14,000 rpm should provide better performance than a conventional Micronair AU4000 atomizer (at 9000 rpm) and a Teejet atomizer for the criterion " numbers of droplets in the 15 to 55 micron range", for the same volume of formulation sprayed. Although the drift mass to optimal droplet mass is twice as high for the SERG/ONAIR as compared to the conventional Micronair AU4000, the ratio of useful numbers of droplets is five times greater for the SERG/ONAIR than for the Micronair AU4000. The expected superiority of the SERG/ONAIR system was investigated in an operational spray comparison of these two atomizers in 1993 in New Brunswick, described below.

TABLE 8.1. Comparison of atomizer performance

	Atomizer Type	Volume Median	Drops/ℓ 15–55μm	Mass Fraction 15–55μm	Mass Fraction < 15 μm	+ Drift Mass / Opt. Mass	≠ Opt. Mass / Waste Mass
1.	SERG/ONAIR with Micronair Sleeve, 14,000 rpm, 2.0 l/min @ 49.2 m/s	34 μm	72 x 10^9	.86	.035	.04	6.4
2.	Micronair AU40000 with 20 Mesh sleeve, 9,000 rpm 7.8 l/min @49.2 m/s	96 μm	14 x 10^9	.25	.004	.02	.34
3.	Tee-jet 11010, 64 m/s, 3.8 l/min	124 μm	11 x 10^9	.17	.009	.05	.21

\+ Defined as Mass Fraction < 15 μm divided by Mass Fraction 15–55μm
≠ Defined as Mass Fraction 15 –55 μm divided by Mass Fraction >55 μm
1. Fluid Properties: Surf. Tens. 30.4 mN/m, Viscosity 16.7 mNs/m^2
2. Fluid Properties: Surf. Tens. 39.8 mN/m, Viscosity 2.3 mNs/m^2
3. Fluid Properties: Surf. Tens. 48.0 mN/m, Viscosity 4.0 mNs/m^2

8.2.2 FIELD CONFIRMATION

In the 1993 spraying program, Forest Protection Ltd. treated 100,000 hectares of spruce-fir forest for protection against the spruce budworm, using fixed-wing spray aircraft [Davies (1993)]. Approximately 28,000 ha were sprayed with the SERG/ONAIR atomizer system mounted on a Dromader M18 aircraft (four atomizers on each wing). Approximately 19,000 ha were sprayed with conventional Micronair AU4000 atomizers mounted on Cessna 188 aircraft (four atomizers each wing). These particular application schemes were set up as a research program to compare the two atomizer systems in terms of eventual protection from defoliation. The program was carried out by the Canadian Forest Service, Maritimes Region, Fredericton, N.B. [Kettela, (1996)]. The pesticide used in both schemes was Futura XLV-HP (a Bt product from Chemagro Ltd.) at two separate applications of 15 BIU/ha, with in all cases a flow rate of 2.0 litres per minute per atomizer. The research program involved extensive pre- and post-spray sampling to determine larval densities and defoliation, and deposit sampling using tree branches themselves. The results are summarized below in Table 8.2.

TABLE 8.2. Summary of spray deposit and reduction of survival and defoliation for Micronair AU4000 and SERG/ONAIR treatment areas (New Brunswick, 1993)

	Atomizer RPM	Deposit, droplets/needle (mean of 3 replications)		Percent Reduction	
		1st Application	2nd Application	Larval Survival	Defoliation
Micronair Au 4000	9300	0.47	0.55	36	31
SERG/ONAIR	14000	0.80	0.87	73	50

The results show that the optimal-droplet atomization system (SERG/ONAIR atomizer system) halved larval survival and reduced defoliation by 61% when compared to the conventional Micronair AU4000. This is operational confirmation of the UULV concept, and provides an explanation of the report [Irving et al., op. cit.] that a conventional Micronair AU4000 operating with a flow rate of 2.0 litres/min/atomizer, at a dosage of 15 BIU/ha, provided approximately the same protection as the same atomizer fed at a flow rate of 6 litres/min/atomizer and at a dosage of 30 BIU/ha. The number of droplets per litre in the preferred size range produced by this atomizer is greater at a flow rate of 2.0 l/min than at 6.0 l/min. The insecticide in that case was Dipel 132 (a Bt product from Abbott Laboratories).

The volume median diameter for the Futura XLV-HP formulation was measured to be 44 microns through the SERG/ONAIR atomizer at 13,500 rpm and 61 microns through the Micronair AU4000 atomizer at 10,000 rpm [Murray (1991)]. These data were obtained from a single "shot" with the Malvern droplet spectrometer through the atomizer spray plume center-line, in the UNB wind tunnel. The values are not total flux based, but are useful for comparison of the two atomizer systems. The data differ from those presented in Table 8.1 because of the very high viscosity, and possibly because of the non-Newtonian character of the Bt formulation, and also because of the different flowrates in the case of the Micronair atomizer.

The SERG/ONAIR system atomizers used in the above spray program were prototypes provided by Effective Particle Technologies Ltd., Fredericton, N.B. This system is not yet commercially available.

8.3 Field Trials for Dose Efficacy Measurements

In the first SERG drift trial in New Brunswick in 1979 [Crabbe et al. 1980)], spray aircraft were repeatedly flown over the same spray line at a few minutes of interval to lay down a succession of line sources of droplets which would be free of interaction of one with the other. This ensured the smoothing out of the effects of random wind fluctuations on the spray cloud far downwind, and on foliage and ground deposits in the near and far downwind field. The source strength was considered to be the sum of the individual sources, and the deposits the sum of the individual deposits. This procedure was also followed in the later Dunphy trials.

Deposit profiles for this type of application are extensively discussed in section 6.4. The basic pattern of foliage deposit downwind of a spray line is a rapid rise to a peak value, followed by a gradual decrease which becomes inversely exponential with increasing downwind distance. Because of this large range in deposit level, this same technique has been used in the field to evaluate pesticide effectiveness, relating pest mortality or foliage protection to pesticide deposit level [Kettela (1993), and Kettela (1990)] as it changes in the downwind direction.

In the second of the above references, the work consisted of a field evaluation of a novel product, Mimic 2F (RH 5992), a growth regulator insecticide produced by Rohm and Haas. The trials involved Forest Protection Ltd. (Fredericton, N.B.), Rohm and Haas Canada Inc., and Natural Resources Canada. Figure 8.1 shows the relationship between spruce budworm larval survival and deposit density (found using a microscope) when the nominal application rate (assuming a track spacing of 30 m) was 1.0 litre/ha of a water suspension formulation providing a dosage of 50 g/ha of active ingredient. This was applied with Micronair AU-4000 atomizers operating at 10,630 rpm at a flow rate of 2.0

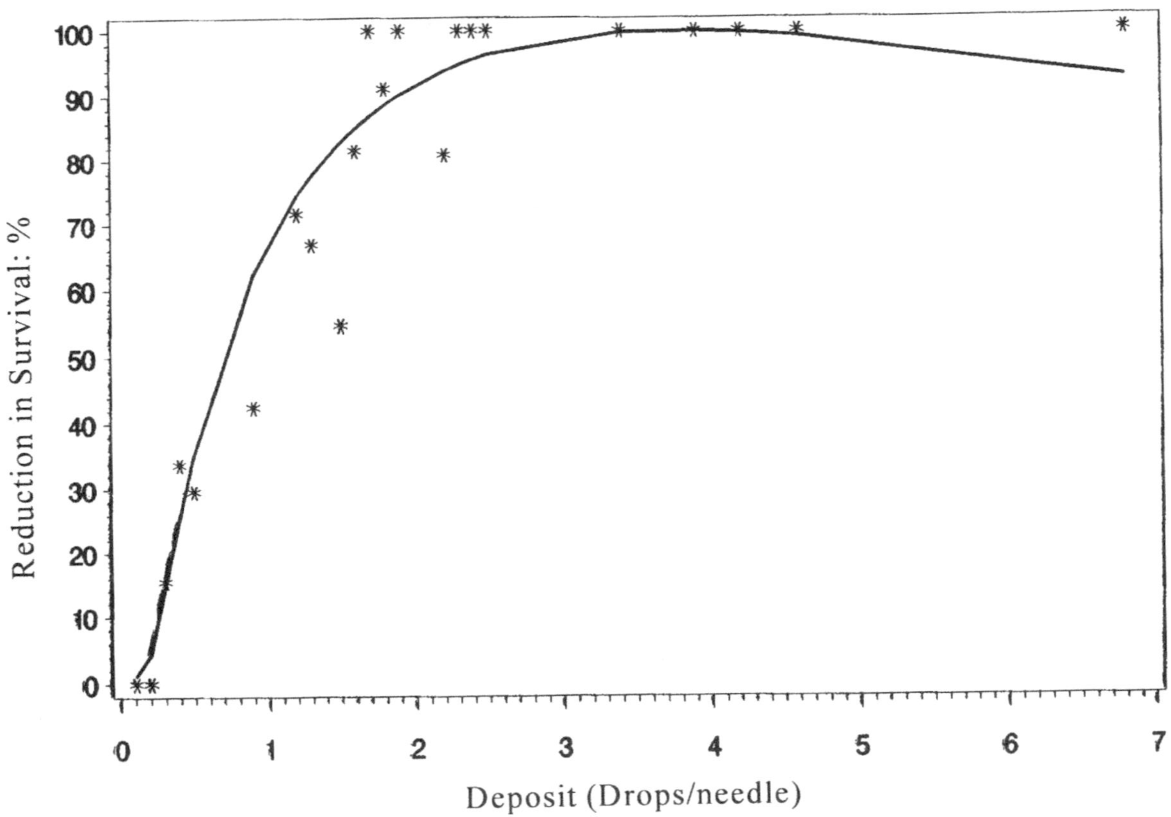

Figure 8.1 Reduction in larval survival as a function of deposited droplets per needle for a 50 g/ha application of Mimic 2F (RH 5992) pesticide against spruce budworm. Solid line is a curve fitted to the experimental data shown. Copied by kind permission of the author [Kettela (1996)].

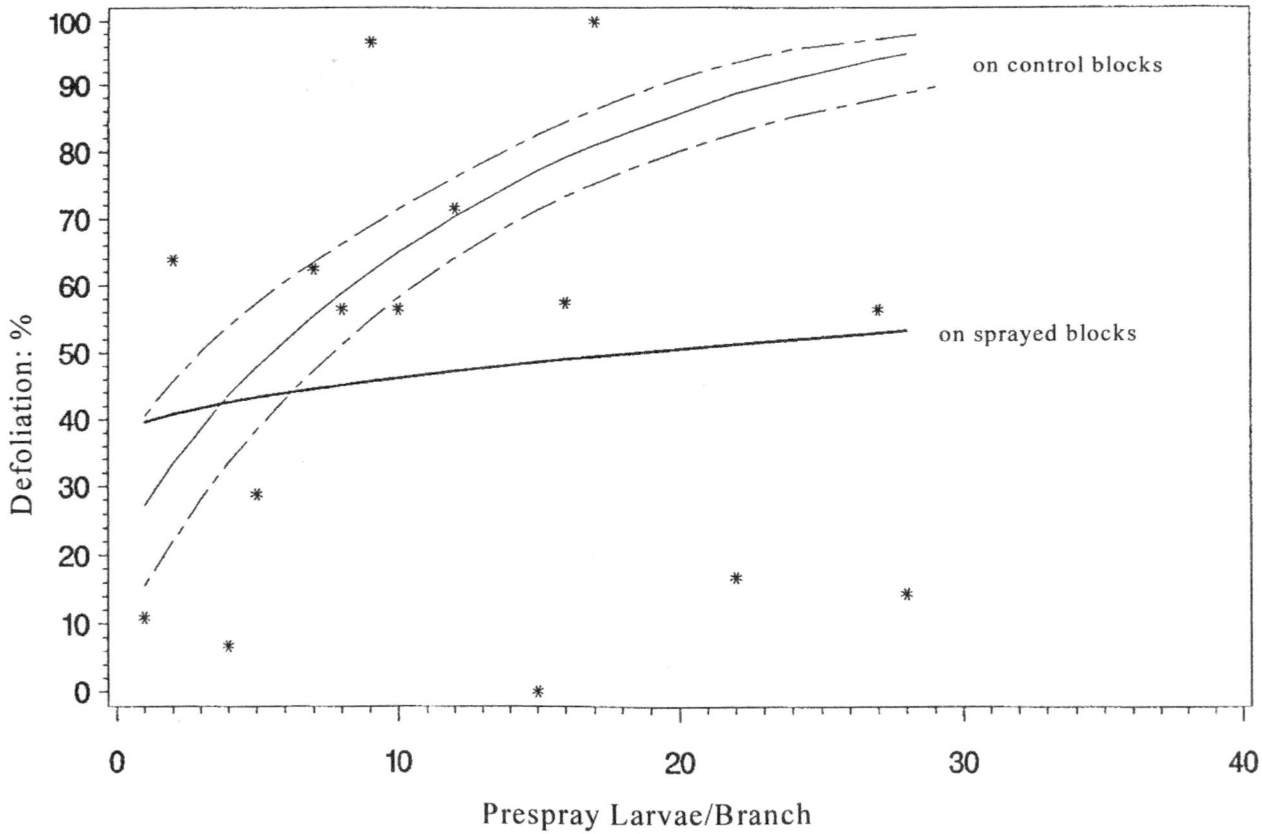

Figure 8.2 Defoliation as a function of prespray larvae per branch, showing defoliation on control blocks with 95% confidence limits indicated. Copied by kind permission of the author [Kettela (1996)].

litre/min., with Cessna 188 aircraft. Figure 8.2 shows the reduction in defoliation for this application. Droplet deposit spectrum was not measured, but was in the size range generally seen for this type of atomizer operation. These and other data presented in the report [Kettela (1993)] were instrumental to the registration of this pesticide for use against spruce budworm in Canada.

Results of laboratory toxicological tests exist for this pesticide on spruce budworm larvae [Nigam and Holmes (1995)]. These were carried out by feeding single uniform-sized droplets to larvae and observing the ensuing mortality. Significant mortality occurred from droplets in the 30 to 40 micron size range and upwards, for a 70 g/ha formulation. These single droplet experiments do not simulate the field situation very well, where a wide range of sizes is seen on the conifer needles. Therefore the field trial method mentioned here appears to be an ideal and even essential stage in the determination of proper pesticide dosage requirements.

8.4 Aircraft Navigation and Data-logging

The Global Positioning System (GPS) of navigation has recently come into use by forestry aerial applicators [Davies (1994)]. GPS technology was developed and set up by the U.S. Department of Defense (DOD) and consists of 24 satellites (21 navigational and 3 active spares) orbiting the earth on a 12-hour cycle on 6 orbital planes at 55° to a polar plane, at 20,200 km altitude. At any time 5 to 8 satellites are visible from any point on the earth [Thistle et al. (1995), Embree (1996)]. The full accuracy of this system is not available to civilian users because DOD has implemented a policy of selectively degrading the signals to reduce accuracy from a 16 meter horizontal positioning accuracy to a 100 to 300 meter accuracy. The civilian community has developed differential GPS, which involves using a GPS receiver at a known location, and using this to correct the position calculated from the satellite signals. The base station corrections can then be used to correct positions calculated by receivers mounted on moving equipment (aircraft, ground-based vehicles, etc.). With differential GPS, 95% of the calculated locations are expected to fall within 10 meters of the true location.

A number of commercial differential GPS systems are available for use with spray aircraft (see Table 8.3). These provide guidance to the pilot to fly directly to a spray block, fly a specific spray pattern, and turn spray booms off and on at the beginning and end of spray lines. A spray plan which is interrupted for any reason can be accurately resumed at a later time. This type of equipment can provide the following advantages [Thistle et al. (op. cit.)]:

1. Help eliminate the problem of treating the wrong area.

2. Provide aircraft tracking and guidance that allows spray material to be applied more uniformly.
3. Supply operational records for quality control and post-operational questions and challenges.
4. Reduce the need for traditional, more costly navigational methods.
5. Minimize lost time in travelling to and from spray plot and airstrip.
6. Improve the ability to respond quickly to unplanned gaps and down-time in the spray application.

TABLE 8.3. Companies marketing aircraft guidance equipment related to aerial spraying [Embree, (op. cit.)].

Company	Brand Name	Location
Agritrack Systems	same	Fall City, WA
AvCan Technologies Inc.	Av Can	Pitt Meadows, BC
Del Norte Technology Inc.	Flying Flagman	Euless, Texas
Picodas Group Inc.	Ag-Nav	Richmond Hill, Ont.
Filanda Inc.	Softnav	Ste-Foy, QC
Precision Electronics Guidance Inc.	Pegs	Warton, TX
Satloc	same	Tempe, Ariz.
Trimble Navigation	Trimflight GPS	Sunnyvale, CA
Wag Corp	Wag Flagger	Tupelo, MS
Zycom Corp.	Agriplan	Bedford MA

The differential GPS equipment all have the following features in common [Embree (op. cit.)]:

1. A GPS receiver with DGPS capability.
2. A light bar equipped with a series of lights which show the pilot his off-track deviation.
3. A moving-map display which shows the pilot the location of the aircraft, the shape of the block to be sprayed, and spray lines and way- points to get to the required locations.
4. Data-logging equipment to store aircraft positions and times as well as other operational variables of the spray apparatus.

For the SERG/ONAIR spray application project described in section **8.2.2**, the M-18 spray aircraft was equipped with a TRIMFLIGHT (Trimble Navigation, New Zealand) real-time data-logging and GPS navigation system. This consisted of a 4000 RLII GPS receiver, a TRIMFLIGHT 486 PC, a radio telemetry link RX (Kenwood TK805D), and a light bar, LCD screen and controller for the spray aircraft; a 4000 RLII GPS receiver and radio telemetry link TX (Kenwood TK805D) for the ground station; a Parrot repeater system with radio telemetry link (Kenwood TK805D) for the spotter aircraft; and antennae and software [Davies (op. cit.)]. The total cost of this equipment was C$88,400. Additional equipment for data-logging purposes were a radar altimeter, sensors for relative humidity, boom pressure, airspeed, and temperature, and a computer control module, for a total additional cost of C$18,100. The repeater system on the spotter aircraft is necessary to transmit differential corrections to the spray aircraft from the ground station when spraying over uneven terrain, since radio communications are line-of-sight at the high frequencies used.

After several startup problems were resolved, the differential GPS system performed exceptionally well in guiding the M-18 pilot to boom on/off way points, from one line to another, and along spray lines. Figure 8.3 shows a typical map showing the actual track of the M-18 over a spray block.

Following upon this initial development, in 1994 Forest Protection Ltd. purchased a AGNAV (Pestechon, USA) real-time data logging/GPS navigating system and installed it in a Cessna 188 Ag Truck . Testing of this and the original TRIMFLIGHT system re-installed in a Bell 206 LR helicopter was carried out and a number of hardware and software "glitches" were experienced, requiring corrective measures.

The USDA Forest Service report by Thistle et al.(op. cit.) describes a differential GPS demonstration project which took place in Montana in October 1994. The systems demonstrated were the AgNav installed in a Cessna AgTruck, and the Satloc system installed in a Ayres Turbo Thrush. The report concludes that this new technology is an important asset to spray application. A number of the conclusions are repeated or paraphrased below:

1. The accuracy of differential GPS systems is equivalent to manufacturers' claims.
2. Independent verification of system coordinates is important as a calibration check.
3. Complex terrain may negate the use of parallel flight lines, requiring additional software development.
4. Detailed pilot training is critical to the correct use of the systems.
5. The systems do not prevent fraud or eliminate the need for observers. However, the number of observers needed is reduced

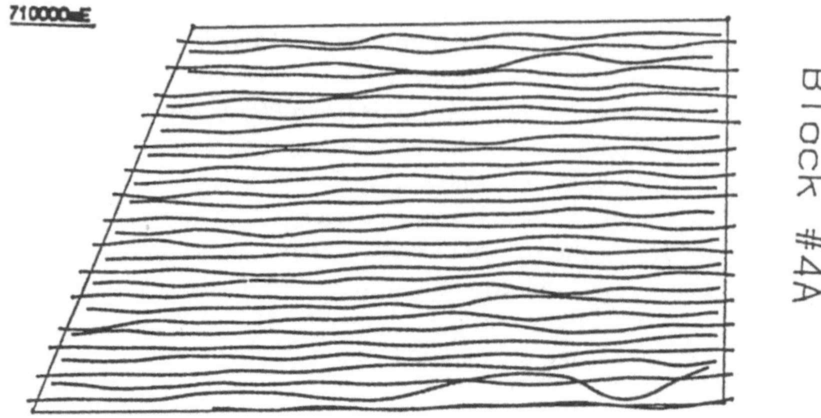

Figure 8.3 Track pattern for Dromader M-18 with GPS-aided navigation by pilot. Block area 2054 ha, track spacing 120 m. Copied from 1993 spray program log by kind permission of D.C. Davies, Managing Director, Forest Protection, Ltd.

It appears that differential GPS will rapidly become an essential tool for forestry aerial spraying.

8.5 Investigating Possible Improvements in Application Methods with PKBW

A validated spray model such as PKBW (or FSCBG, or AGDISP) can be used to test application situations at the cost of a few minutes of computer time. This is illustrated below with a herbicide application problem and an insecticide application problem for conifer defoliation control.

8.5.1 UTILIZATION AND DRIFT OF A PHYTOTOXIC HERBICIDE

We consider here the herbiciding of a recently-planted plot with seedlings placed at a two-meter spacing. The plot is 500 meters upwind of a conifer stand shelterbelt (a Dunphy type stand (see section 2.1) 100 meters deep, beyond which there is an area of vegetation which we assume may be sensitive to the phytotoxic material to be sprayed). The spray application is to be done by a Bell 206 helicopter flying at 31.3 m/s (70 mph). The spray nozzles of choice are the CP nozzles using the 0.125 inch orifice and the 45° deflector plate, with the assembly facing towards the rear of the aircraft (see droplet spectrum data in Appendix II). There are 26 nozzles distributed along a boom which extends 1.4 meters beyond the end of the rotor span. The flow rate per atomizer will be 6.0 litres/min. of a glyphosate formulation which contains 3% by volume of active ingredient dissolved in water. We assume the formulation evaporates as water, with the 3%

glyphosate being non-volatile. These application parameters are similar to those for typical herbicide applications in New Brunswick [Riley et al.(1991)]. The PKBW model inputs are summarized in Table 8.4 below, and the deposit results for a first swath applied 500 m upwind of the shelterbelt are given on Figure 8.4, with the results for the last swath applied at 100 m upwind of the shelterbelt given on Figure 8.5.

TABLE 8.4 Summary of input data for the PKBW model

SIMULATION OF BOOKCH8	CASE NO. 1	BELL 206	STABLE	
DATE OF SIMULATION:	22 6 1996			
RELATED OUTPUT FILES: CH8.SUS	CH8.CHT			
AIRCRAFT AND ATOMIZER PARAMETERS:				
AIRCRAFT TYPE=	BELL 206	ROTOR DIAM.=		10.2 m
AIRCRAFT SPEED =	31.3 m/s	AIRCRAFT HEIGHT =		15.0 m
SPREAD UNDER THE ROTOR =	13.0 m	DISTANCE FROM FUSELAGE =		.0 m
ATOMIZER TYPE =	OTHER	FLOW PER ATOMIZER =	6.01	/ min
NUMBER OF ATOMIZERS =	26	EMSN. OF A.I. OR TRACER =		2492 mg/m

DROPLET SPECTRUM VOLUME FRACTION	DIAMETER (μm) (Nominal)	(BEFORE EVAPORATION) (Range)
.01	20.00	0-40
.03	60.00	40-80
.09	110.00	80-140
.23	200.00	140-260
.24	310.00	260-360
.21	510.00	360-460
.19	730.00	460-1000

VORTEX TIME	VORTEX UPWIND CIRC.	VORTEX DOWNWIND CIRC.
.00	368.82	351.52
1.00	295.67	189.77
2.00	112.71	77.56
3.00	69.19	45.92
4.00	53.42	24.07
5.00	30.90	7.32
6.00	13.37	1.38
7.00	8.52	.00
8.00	2.78	.00
9.00	.00	.00

WEATHER PARAMETERS:	
ATMOSPHERIC STABILITY CLASS =	STABLE

AIR TEMPERATURE = 15.00 Deg C		AIR PRESSURE = 101.00 kPa	
RELATIVE HUMIDITY = .40		MONIN-OBUKHOV LENGTH = 800.00 m	
STAND		FRICTION VELOCITY (m/s)	MIXING LAYER HIT (m)
1.		.200	250.00
2.		.200	250.00
3.		.200	250.00
STAND PARAMETERS:			
STAND NUMBER 1.			
STAND TYPE =	SHEEPHOUSE BK. CONIFER PLANTATION EXTENDING 500 m DOWNWIND OF FIRST SPRAYLINE:		
ROUGHNESS HT. =	.23 m	BASE OF THE FOL. CROWN=	.50 m
MEAN TREE HT. =	1.62 m		
NEEDLE OR LEAF D. =	87 mm	NEEDLE OR LEAF L. =	87 mm
FOLIAGE DEN. =	.40 SQUARE METERS OF FOLIAGE/SQUARE METERS OF GROUND (BROADLEAF ONLY, CONIFER SEEDLINGS IGNORED)		
STAND NUMBER 2.			
STAND TYPE =	DUNPHY MIXED CONIFER SHELTER BELT EXTENDING 100 m DOWNWIND OF STAND 1		
ROUGHNESS HT. =	1.44 m	BASE OF THE FOL. CROWN =	7.50 m
MEAN TREE HT. =	15.00 m		
NEEDLE OR LEAF D.=	.20E+01 mm	NEEDLE OR LEAF L. =	10E+02 mm
FOLIAGE DEN.=	2.94 SQUARE METERS OF FOLIAGE/SQUARE METERS OF GROUND		
STAND NUMBER 3.	The same foliage density and distribution of a phytoxically sensitive plant species as in stand number 1 above, extending 280 m downwind of shelter belt.		
FORMULATION PARAMETERS:			
CARRIER =	Water		
MOL. WT. OF CARRIER =	18.00 g/mole	DEN. OF CARRIER=	10E+04 Kg/m^3

The deposit results can be summarized as follows in terms of % of applied active ingredient:

	DEPOSITS			DRIFT
	Ground	Broadleaf	Conifer#	Beyond shelterbelt
1st swath	83.1%	8.1%	0.7%	8.1%
last swath*	81.7%+	8.1%	3.1%	8.0%

\# Within shelterbelt
* Adds up to more than 100% because some drift beyond shelterbelt deposits downwind.
\+ 3.2% within shelterbelt and beyond

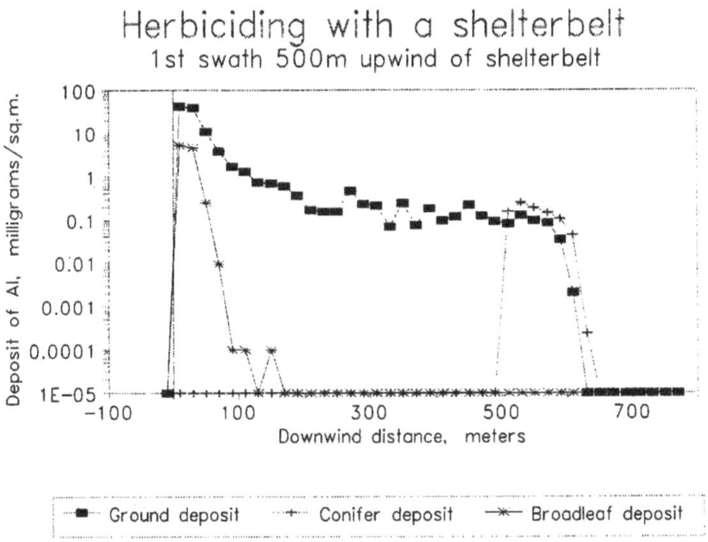

Figure 8.4 Deposit pattern for spray plot of Table 8.4. A background deposit of 10^{-5} mg/m^2 of AI is assumed to permit use of the logarithmic scale. Shelterbelt conifers from 500 to 600 m.

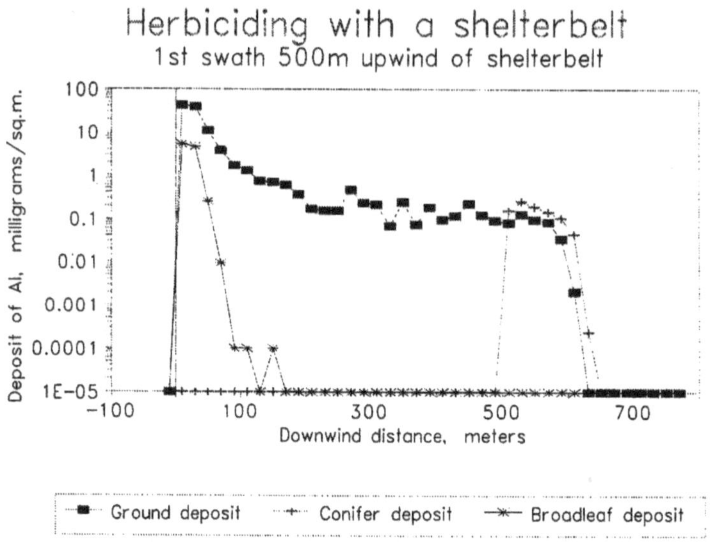

Figure 8.5 Results as above for the last spray swath applied 100 m upwind of the edge of the conifer shelterbelt stand.

An examination of these results shows the following:

1. Only about 10% of the herbicide reaches its planned destination. Most of the spray is found on the ground.
2. The conifer shelterbelt removes a significant portion but not all of the droplets drifting into it.
3. The simulation can be used to plan alternatives in the application method.

The problem of low utilization of the herbicide is a direct result of the fact that the vegetation is sparse in this case of conifer release. The Sheephouse Brook, N.B. plot [47°05'N, 66° 06'W ; Riley (1992)] was scheduled for actual treatment a year after stand characteristics were measured [Chapman (1993)], so that the characteristics used for foliage density are perhaps 25% lower than for a typical application. In the typical situation modelled, about 12% of the herbicide would reach its planned destination. Note that leaf orientation is not always perpendicular to the droplet velocity vector, and this reduces the interception of droplets by the leaf. This effect is incorporated in PKBW through the simulator/foliage calibration factors. [The field trial results reported by Riley (op. cit.) gave a value of 13% for the integrated canopy deposit to 300 m for application with a Thru-Valve Boom atomizer with a similar spectrum to that used in the simulation ($D_{v.5}$ = 311µm)].

Drift could be reduced by flying the aircraft at lower speed (generating larger droplets) near the shelterbelt upwind edge, but this would not affect the herbicide utilization factor significantly. It may be desirable to consider the relative merits of using a systemic herbicide which would be taken up by the roots of the broadleaf plants. This is not a new concept by any means. What we illustrate here is a new method of assessing application methods which is extremely effective.

8.5.2 "SMART" FORMULATIONS

We define here a "smart" formulation as a wettable powder [Sundaram (1995)], a solid particle coated with a non-soluble insecticide suspended in a non-toxic liquid, water, for example. The solid particle has a controlled size and the insecticide is of sufficient potency to make it effective against the target insect at levels of one particle per needle on conifers. It is necessary that the formulation be atomized so that the powder separates from the carrier liquid. It appears from our experience that a high-energy atomizer such as the SERG/ONAIR can achieve a very good separation of individual particles. Modelling can be used to assess the deposit and drift behaviour of such a formulation. Using the conditions of Table

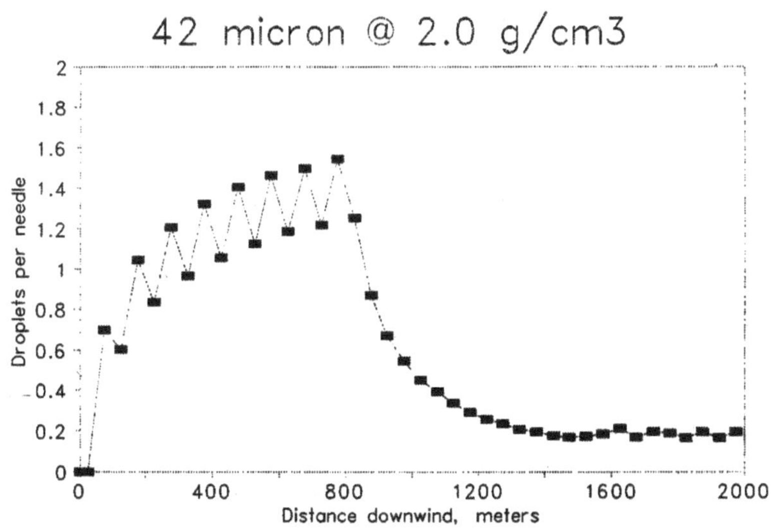

Figure 8.6 Needle deposit for an eight-swath application of wettable powder in water suspension at 25% solids concentration. Ideal conditions of particle separation assumed. Dromader M-18 aircraft with 8 SERG/ONAIR type atomizers, total flow rate 16 litres/min, flying speed 47 m/s, height 35 m, track spacing 100 m, unstable atmosphere at $u^*=0.51$ m/s, L=-42 m, Dunphy forest.

6.3, for an application over a Dunphy-type stand using Dromader aircraft, Figure 8.6 shows the result of spraying with 42μm solid particles of density 2.0 g/cm^3 suspended in water at a 25% concentration. The particle size and density are rough approximations for a mineral particle (limestone, dolomite, or other relatively heavy particle) coated with insecticide.

The drift characteristic of the 42 micron particle at a density of 2.0 g/cm^3 is identical to that of 60 micron droplets at 1.0 g/cm^3 (see Fig. 6.8 b)); 25% of the particles drift beyond the 2000 m downwind position. By comparison, drift of 15 micron droplets at 1.0 g/cm^3 in the same conditions amounts to about 60% of the applied spray.

Significant drift reduction appears achievable with the wettable powder approach. The practical problems which require solution are:

1. Can an appropriate formulation be manufactured within the constraints specified?
2. Does the presence of the particle interfere with the effectiveness of the pesticide?
3. Will the available atomization equipment be efficient enough to separate the particles from the carrier liquid?
4. Will the insecticide remain on the solid particles?
5. Can an effective product be made at a reasonable cost ?
6. Will the atomizer be subject to unreasonable erosion by the particles?

These questions are well understood by formulation chemists, who already have made significant advances in the field of wettable powders. We demonstrate above the utility of a validated model in answering a number of application technology problems such as 1)required flow rates to achieve adequate coverage, 2)expected drift, and 3)expected atomization requirements, i.e. perfect separation of the particles.

References

Chapman, T.J. (1993): Modelling of Herbicide Spray Dispersion and Deposition, M.Sc. Thesis, UNB Dept. of Chemical Engineering, Fredericton, N.B.

Crabbe, R.S., L.Elias, M. Krzymien, and S. Davie (1980): New Brunswick Forestry Spray Operations: Field Study of the Effect of Atmospheric Stability on Long-Range Pesticide Drift, report #LTR-UA-52, National Research Council, National Aeronautical Establishment, Ottawa, Can.

Davies, D.C. (1994): Some Operational Parameters for Consideration in Aerial Pesticide Applications and their Potential Use for Future Development, in Capco Technical Session: Overview of Spray Drift and Deposit, W. Sexsmith ed., Natural Resources Canada Can. Forest Service, Science and Sustainable Development, Hull, Quebec., pp 53-75.

Davies, D.C. (1993): 1993 Spruce Budworm and Hemlock Looper Aerial Treatment Program Report, Forest Protection Ltd, Fredericton, N.B., Aug 6, 1993, 31 pages incl. app.

Embree, D.G. (1996): Capabilities of Global Positioning System and Data Logging Equipment to Guide Track and Record Application History of Aircraft and Ground Sprayers in Forest Operations, report to SERG, available at Canadian Institute for Scientific and Technical Information, Ottawa, 28 pages.

Himel, C.M. and A.D. Moore (1969): Spray droplet size in the control of spruce budworm, boll weevil, bollworm, cabbage looper, J. Econ. Ent., **11**, pp 916-918.

Irving, H.J., E.G. Kettela, and J.J.C. Picot (1990): Technology transfer: An example, Proc. ACAFA Symposium and Trade Show, Winnipeg Man., NRC Canada, Ottawa, pp 133-139.

Kettela, E.G. (1996): B.t. Optimum Use Trials: Evaluation of the SERG/ONAIR spray system in a pilot field application; report to SERG, available from the Canadian Institute for Scientific and Technical Information, Ottawa.

Kettela, E.G.(1993): Evaluation of RH 5992 (Mimic 2F) Aerial Spray Trials Against Spruce Budworm In New Brunswick in 1992, report to SERG, available at Canadian Institute of Scientific and Technical Information, Ottawa, 51 pages incl. app.

Kettela, E.G., and V. Steel (1990): An Account of Spray Trials Conducted to Evaluate the Efficacy of B.t. against High Spruce Budworm Populations. 1990; report to SERG, available at Canadian Institute of Scientific and Technical Information, Ottawa, 72 pages incl. app.

Murray, D. (1991): An Analysis of Bt Sprays and Simulants, UNB research report, Department of Chemical Engineering, Fredericton, N.B., Aug. 29.

Nigam, P.C., and S.E. Holmes (1995): Determination of Optimum Drop Sizes of RH 5992 (Mimic 2F) against Spruce Budworm Larvae for Stomach and Contact Toxicity, report to SERG, available at Canadian Institute of Scientific and Technical Information, Ottawa, 35 pages.

Picot, J.J.C. (1990): The SERG/ONAIR high rpm atomizer, Proc. ACAFA Symposium and Trade Show, Winnipeg, Man., NRC Canada, Ottawa, pp 191-194.

Riley, C.M., C.J. Wiesner, and W.A. Sexsmith (1991): Estimating off-target spray deposition on the ground following the aerial application of glyphosate for conifer release in New Brunswick, J. Environ. Sci. Health, **B26**, pp 185- 208.

Sundaram, A. (1995): Pesticide Formulations, chapter 46 in Forest Insect Pests in Canada, J.A., Armstrong and W.G.H. Ives, eds., Natural Resources Canada, Can. Forest Service, Science and Sustainable Development Directorate, Ottawa, pp 453-464.

Thistle, H., A. Jasumback, W. Kilroy, K. Mierzejewski, and J. Barry (1995): DGPS in Aerial Spraying in Forestry: Demonstration and Testing, Final Report, #3400-FPM, USDA Forest Service Technology and Development Program, Missoula, Montana.

van Frankenhuyzen, K. (1995): Development and current status of Bacillus thuringiensis for control of defoliating forest insects, in Forest Insect Pests in Canada, J.A. Armstrong and W.G.H. Ives, eds., Natural Resources Can., Can. For. Serv., Science and Sustainable Dev. Directorate, Ottawa, pp 315-325.

CHAPTER 9

PESTICIDES AND GOVERNMENT REGULATIONS

9.1 Introduction

The marketing of pesticides for use, and the aerial application of pesticides are usually strictly controlled in developed countries. To illustrate this, we repeat here the "Part II: Purpose" section of the The New Zealand Resource Management Act 1991, which gives a sense of the direction taken by recent public policies in that country:

> (1) The purpose of this Act is to promote the sustainable development of natural and physical resources.
>
> (2) In this Act, "sustainable management" means managing the use, development, and protection of natural and physical resources in a way, or at a rate, which enables people and communities to provide for their social, economic, and cultural well being and for their health and safety while-
>
> (a) Sustaining the potential of natural and physical resources (excluding minerals) to meet the reasonably foreseeable needs of future generations; and
>
> (b) Safeguarding the life-supporting capacity of air, water, soil, and ecosystems; and
>
> (c) Avoiding, remedying, or mitigating any adverse effects of activities on the environment.

In response to this legislation, individual regional and district Councils in New Zealand have been developing detailed regulations for pesticide use in their jurisdictions [Richardson (1996)] which, presumably, will result in significant changes in the regulatory as well as the physical environments where these products are used.

In view of the fact that this text is concerned with new developments in pesticide application methods, we supply some information here on regulations in effect in Canada and the USA. It is important to note that these regulations change with time, and are given here to provide a context in which forestry spraying is conducted. Both countries have set up special studies regarding the assessment of pesticide drift. In Canada, the Canadian Interdepartmental Task Force on Pesticide Drift was set up in 1989 to develop new regulatory guidelines regarding pesticides using a scientific approach [Nicholson (1994); Mickle, (1996)]. In the USA, the Spray Drift Task Force was set up with representatives from the agricultural chemical industries to achieve similar ends. Both of these entities are in the process of carrying out their mandates, and progress in the form of changes in spray practice and/or regulations should soon appear.

We restrict ourselves here to 1) an overview of federal and provincial regulations in Canada, and 2) a simplified review of the the U.S Federal Insecticide, Fungicide, and Rodenticide Act (commonly called FIFRA), and 3), a summary of a survey of US and European programs for training pesticide applicators and inspecting application equipment.

9.2 Federal and Provincial Regulations in Canada

We summarize here a document prepared by Mr. H.J. Irving, Coordinator of SERG, formerly Managing Director of Forest Protection Ltd, Fredericton, N.B., who has kindly supplied it for our use [Irving (1993)]. It is noted that since the subject matter is constantly evolving, the reader is well advised to verify the existing situation at any specific time:

AIRCRAFT USE AND REGULATION FOR PESTICIDE APPLICATION IN CANADA
by

H.J Irving

June 29, 1993

INTRODUCTION

In Canada there are two levels of government, federal and provincial, and many pieces of legislation at each level that play a role in the regulation of aerial application of pesticides.

Transport Canada has jurisdiction over the airworthiness of the aircraft and the pilot's ability to fly that type of aircraft.

Agriculture Canada through the Pest Control Products (PCP) Act has the authority to regulate pesticides used in agriculture, and also those used in forestry, industry, and public health. This Pest Control Products Act requires that every pesticide product is registered by Agriculture Canada before it can be sold or used in Canada. Before registration is granted, the manufacturer must provide scientific data which show that the pesticide is effective for the uses indicated on the label and that it is safe when used as directed. When registration has been approved, the product is assigned a Pest Control Products Act Registration Number that must appear on the label and is specific to that product for all time. Applicators must not use pesticides for purposes other than those described on the label, unless otherwise specifically authorized by provincial and federal authorities.

In addition to the Pest Control Products Act, a Number of other federal acts may have implications limiting the use of pesticides, including:

Food and Drug Act
Canadian Environmental Protection Act
Fisheries Act
Migratory Birds Convention Act
Canada Water Act
Transportation of Dangerous Goods Act

Detailed examination of these acts is beyond the intent of this review. Neither is this an exhaustive list of federal legislation having implications relative to pesticide use.

All ten provinces have enacted legislation regulating how pesticides may be handled and applied. Such legislation does not contradict federal legislation - in cases where two regulations cover the same area whichever is more restrictive applies.

Although the provincial acts are not identical they tend to follow the same pattern - they in a general way authorize a minister of the government to make regulations governing pesticide use. These provincial regulations indicate to what extent aerial application is controlled and monitored for mitigation of risk to non target organisms and the environment in general. An overview of these regulations is one of the major objectives of this exercise.

Printed regulations by themselves are not indicative of their application or enforcement. For this reason the author visited the appropriate officials of six provinces and discussed the intent and application of their regulations. Through telephone conversations, officials of the other provinces explained how their regulations are applied.

2. PROVINCIAL LICENSING AND CERTIFICATION

Regulations and terminology regarding licensing of aerial applicators of pesticides vary from province to province but all require that the applicator be licensed and that evidence of adequate knowledge be provided (usually through a written examination) concerning the use, application, handling and storage of pesticides.

Some provinces have two-tiered system of licensing where both the business operator and the pilot each have a license or certificate. Several require a basic applicator's license for a specific purpose such as agriculture, forest tending or biting fly control and in addition another one for the use or application of pesticides from an aircraft. Provincial requirements for aerial applicator's licenses are included in Appendix A.

3. PERMITS

Most provinces require some form of permits for aerial treatment near sensitive areas whether because of susceptible ecology or proximity to humans. These permits usually call for special precautions to reduce risk to sensitive areas and organisms with special attention to drift avoidance. Provincial policies are outlined in Appendix B.

4. AIRCRAFT SUITABLE FOR PESTICIDE APPLICATION

(The aircraft registration listings maintained by Transport Canada (federal government ministry) can be consulted for aircraft of the type normally used in aerial application. Spray aircraft are not listed separately, nor are aircraft which are temporarily converted to spray use).

5. AIRCRAFT EQUIPMENT

Neither federal nor provincial regulations specify standards for dispersing equipment and only seldom is there any mention of such equipment on the permits issued by provincial regulators.

6. BUFFER ZONES

Most provinces specify certain types of sensitive areas that could be negatively impacted by pesticide trespass from adjacent spray operations and which must be protected from pesticide drift.

A buffer zone is a strip of land between the sensitive area and the treatment area. Pesticides are not applied directly to the buffer zone, but it must be wide enough to contain all pesticide residue drift or runoff in the direction of the sensitive area.

Field experience has shown extreme variation in the distance pesticides can drift during "normal" aerial applications. This emphasizes the difficulty of determining a scientifically based numerical width for a buffer zone for regulatory purposes. Some provinces have included mandatory buffer zone widths in their regulations and/or permits while others have issued only guidelines and charge the applicator with the responsibility for avoiding pesticide trespass into these sensitive areas.

A summary of the policies being followed by provinces is contained in Appendix C.

8. BITING FLY CONTROL:

Aerial application of chemical pesticides for the control of black flies and mosquitos was quite common some years ago. More recently this practice has been used less and less. This season (1993) only Manitoba has authorized the aerial application of chemical pesticides for mosquito control.

9. DRIFT REDUCTION AND MITIGATION OF RISK

The difficulty of reducing risks through licensing, permits and buffer zones is highlighted by the variety of different policies being implemented in the various provinces. Buffer zones in particular are extremely difficult to quantify scientifically since weather variations at the time of application and small differences in aircraft equipment can result in major differences of deposition away from the aircraft track. Moreover, most buffer zone width requirements do not specify wind direction relative to the sensitive area. The variations from province to province illustrate the questionable scientific basis for these regulations.

Provinces have attempted to reduce risks of drift through various guidelines and recommendations. Examples are: a) in Prince Edward Island "the last 30 meters of a field bordering roads, highways, occupied habitations and streams must be sprayed parallel" and b) in British Columbia "begin the first spray swath, whenever possible, in an upstream direction along the border of the proposed treatment area".

The efforts of the Canadian Aerial Applicators Association in carrying out aircraft calibration clinics were noted. These clinics measure the deposit patterns from the aircraft through actually flying over a sensitive receptor strip or string which can be quickly analyzed for amounts of tracer which has been incorporated in the simulated pesticide. Areas of too high and too low deposit are then adjusted through changing nozzle numbers and spacing to provide a more even distribution pattern. Nozzle orientation and tip condition are checked. End nozzles on the spray boom are not placed further out than three-fourths of the wing length and at least three feet inboard from the wing tip. This is to prevent small droplets from becoming entrapped in the wing tip vortices and being carried aloft - an important factor in reducing drift. The test is then repeated to verify an acceptable swath width and deposit profile. Further adjustment and retesting is carried out as may be necessary.

Attendance at these calibration clinics is entirely voluntary and unfortunately it seems that only the more conscientious operators participate. British Columbia (officialdom) is wondering if annual calibration of spray aircraft via such a procedure should be mandatory.

The initiatives of the Canadian Association of Pesticide Control Officials (CAPCO) in developing national core standards has been enthusiastically received in all provinces. In discussing their current practices most emphasized that they were in process of changing their training materials and regulations and that these changes would reflect the concepts from the national core standards and the aerial application module.

All those interviewed agreed that the greatest improvement in mitigation of risk from aerial application would be achieved through training and education of the applicators.

APPENDIX A

PROVINCIAL REQUIREMENTS FOR AERIAL APPLICATOR'S LICENSES

BRITISH COLUMBIA:
Aerial Applicator's Licenses are issued under the authority of the B.C. Pesticide Control Act and are administered by B.C. Environment. British Columbia has a two-tiered system requiring the operator to have a "Service License" and who must be, or employ, a pilot whose "Applicator Certificate" is endorsed with the appropriate category - agriculture, forestry, or mosquito and biting fly control, etc.

ALBERTA:
Aerial Applicator's Licenses are issued under the authority of the Alberta Agricultural Chemicals Act and are administered by Alberta Environment and require:

Class A License - Agriculture + Class E - Aerial application
Class L License - Biting flies + Class E - Aerial application & etc.

Alberta's regulations are in process of changing and will soon involve a two-tiered system with service operators who must employ licensed applicators somewhat similar to British Columbia.

SASKATCHEWAN:
Applicator permits are issued under the authority of the Pest Control Products Act and administered by Saskatchewan Agriculture:

Permit Class 1A1: Commercial Air for agricultural and forest lands
Permit Class 1A2: Flying Farmer for agricultural and forest lands & etc.

To obtain a permit an applicant must pass a pesticide applicator's course offered by the Saskatchewan Institute of Applied Science and Technology Saskatchewan intends to introduce a two-tiered system including both Service Licenses and Applicator Permits.

MANITOBA:
Aerial Applicator Licenses are authorized through the Manitoba Pesticides and Fertilizers Control Act and administered by Manitoba Agriculture:
Class II License plus a Commercial Pilots License or a Flying Farmer Exemption from the Federal Air Transport Committee.

ONTARIO:
Aerial Applicator Licenses are authorized through the Ontario Pesticides Act and administered by the Ontario Ministry of the Environment.
Ontario has a two-tiered licensing system - Licensed Operators who must have Licensed Exterminator(s) on staff. For aerial application the pilot must have:

Exterminator's License Class 7 for pesticides other than herbicides.
Exterminator's License Class 8 for herbicides.

QUEBEC:
Aerial Application Certificates are authorized through the Quebec Pesticides Act and administered by the Ministere de l' Environnement:

Certificate Class "C.1": Aerial Application

NEW BRUNSWICK:
Pesticide Operator's Licenses and Pesticide Applier's Certificates are issued under the authority of the New Bruswick Pesticides Control Act and administered by the Department of the Environment. New Brunswick has a two-tiered system requiring:

Pesticide Operator's License for the operation of the business and Pesticide Applicator's Certificates:

Class A1 - Aerial - with/without spotter supervision
Class A2 - Aerial - with spotter supervision
Class A3 - Aerial - supervision (spotter)

NOVA SCOTIA:
Business Operator's Licenses and Aerial Applicator's Licenses are issued under the authority of the Pest Control Products Act and administered by the N.S. Department of the Environment.
Nova Scotia has a two-tiered system requiring a Class VII - Business Operator's License which operator is/or has on staff a person who holds a Class VI - Aerial Applicators License.

PRINCE EDWARD ISLAND:
"Class B" Aerial Applicator's Licenses are issued under the authority of the P.E.I. Pesticides Control Act and administered by the P.E.I. Department of Agriculture.

NEWFOUNDLAND:
Aerial Applicator's Licenses are issued under the authority of the Newfoundland Pesticides Control Act and administered by the Department of Environment and Lands and require:

Class 1: Forest tending + Class 5: From an aircraft
Class 2: Biting fly control + Class 5: from an aircraft & etc.

APPENDIX B

PROVINCIAL POLICIES REGARDING PERMITS FOR AERIAL APPLICATION

BRITISH COLUMBIA:
Permits are administered by the Ministry of Environment. In addition to the federal classification of pesticides "restricted", "commercial" and "domestic", British Columbia regulations define two other pesticide schedules which are "permit-restricted" and"exempted". The provincial requirements for storing, selling or applying pesticides are different for each schedule.
"Permit-restricted" pesticides include the persistent organochlorines such as chlordane, dieldrin, endrin and heptachlor and highly toxic pesticides such as aldicarb and calcium cyanide.Purchase and use of these of this category is restricted to certified applicators possessing a "restricted permit".

For federally classified "restricted" and "commercial" pesticides a "Pesticide Use Permit" must be obtained for use on public land and private land used for forestry , transportation or public utility purposes.
Applications for Pesticide Use Permits are reviewed by an inter-ministry committee for hazards to human health and the environment. When hazards are identified, a permit may be denied or issued subject to specific conditions. It is the applicator's responsibility to ensure that all permit conditions are strictly followed, including the treatment location, treatment area, pesticide product, application rate and method, pesticide free zones and public notification.
There is a special requirement for aerial application of pesticides. Where aerial application is not under a permit (i.e., on private land), written notification of a proposed application must be sent to the nearest Pesticide Management Program Regional Manager. The notification must be in writing(i.e., fax or letter) and at least 48 hours before application.

ALBERTA:
Permits from the Minister Of Environment are required for application of pesticides to:

a) an open body of water or within a horizontal distance of 30 meters from an open body of water.
b) an area in forestry production on public land.
c) any area for the control of vegetation along any highway, road, railroad, pipeline, utility line or telephone line, or on grazing leases on public land or on any other public lands in any Green Area of the province

Specifically permits are required for forestry budworm spraying and will only be issued for B.t. and there will be no chemical herbicide spraying for forestry in the "Green Area". The use of "Vision" has been classed as experimental to date.
The maximum use of forestry herbicides (although classed as experimental) will not exceed 2500 hectares in 1993.
Applications for permits are reviewed and decisions taken in a manner not dissimilar to that described for British Columbia.

SASKATCHEWAN:
Saskatchewan does not require specific permits for aerial application for agricultural or forestry purposes. Budworm spraying does not require a permit. However, permits are required for applications of pesticides to water and for biting fly control. These permits are processed through the Department of the Environment.

MANITOBA:
Permits are required for application of pesticides to:

a) Bodies of water.
b) Private property to which the public has access.
c) Any areas for or on behalf of government departments, crown corporations, municipal corporations, etc. or on crown lands or rights of way on behalf of private corporations.

Most common permit requirements are for forestry insecticides and herbicides and for biting fly control. These permits are administered under the authority of the Environmental Act.

ONTARIO:
The Ministry of Environment has classified pesticides into six categories principally based on their toxicity and persistence. Permits are required for aerial application:

1) when applying a Schedule 1 or 5 pesticide or a Schedule 2 pesticide containing a hormone type herbicide or TBA, fenac, picloran or paraquet; and
2) when performing an extermination with the objective of improving production of Crown timber.

On private land, if the conditions mentioned in 1 and 2 above do not apply , a permit is not required. Therefore, in the case of private land forest insect control programs (i.e. gypsy moth control in cottage country) a permit is not required.
However, aerial application for larviciding biting flies is considered an aquatic extermination and the applicator requires the appropriate Water Class exterminator's license and permit.

QUEBEC:
The Pesticides Act Regulations classifies pesticides into five categories. Based on this classification:

1) a Commercial Use Permit is required for aerial application by a person who performs for others, for remuneration, work involving the use of pesticides in classes 1 to 4; and
2) a Private Use Permit is required for aerial application by a person who performs, other than through a permit holder, work involving the use of pesticides in Classes 1 to 3.

NEW BRUNSWICK:
All aerial applications of pesticides require permits.

NOVA SCOTIA:
All aerial applications of pesticides require permits.

PRINCE EDWARD ISLAND:
All aerial applications of pesticides require permits.

NEWFOUNDLAND:
Aerial applications of pesticides require "Operator's Licenses" issued to the operating company. These are somewhat equivalent to what are considered permits in most other provinces.

APPENDIX C

PROVINCIAL POLICIES REGARDING BUFFER ZONES

BRITISH COLUMBIA:
Pesticide Use Permits specify a "10 meter pesticide free zone" along and around all bodies of water. Buffer zones are normally recommended rather than enforced. Recommended buffer zones in forestry are 100 meters for helicopters and 200 meters for fixed wing aircraft. These recommended buffer zones are in addition to the 10 meter pesticide free zone.

ALBERTA:
Regulations prohibit the aerial application of pesticides within 30 meters of an open body of water. Otherwise Alberta does not stipulate buffer zones for agriculture. Permits are required for aerial forestry applications and may include whatever buffer zones, exemptions or restrictions considered appropriate for the particular situation.

SASKATCHEWAN:
Aerial application of pesticides within 50 meters of of the high-water level of surface water or a watercourse. Other buffer zones are not designated in Saskatchewan.

MANITOBA:
Aerial application of pesticides within 30 meters of open water is prohibited without a permit for such applicaton. Permits for forestry herbicide application require a 30 meter buffer zone. No buffer zones are enforced for B.t. application for forestry budworm control. Permits for mosquito control prohibit the spraying of adulticides to parks, playgrounds or ball diamonds when occupied by the public and also prohibit the spraying of insecticides over private residences.

ONTARIO:
Buffer zone setbacks to protect sensitive adjacent areas are not prescribed by regulation. However, they are included in permits for aerial application of pesticides in CROWN FORESTS in meters called guidelines as follows:

DESIGNATED AREAS	HERBICIDES	CHEMICAL INSECTICIDES
Fish bearing waters	60 m	120 m
Ecological reserves	120 m	240 m
Human habitations	120 m	240 m

No buffer Zones are specified for B.t.

Numerically based buffer zones are not imposed for agricultural aerial spraying. The policy of the Ministry of Environment is that it is the responsibility of the applicator to assess each situation individually to determine appropriate buffer zones.

QUEBEC:
No buffer zones are specified for agricultural applications.
For aerial applications to forests a 60 meter setback from waterways and occupied habitations is required. Since all forestry applications are by permit only, each case is assessed individually and appropriate buffer zone requirements are imposed as the situation may warrant.

NEW BRUNSWICK:
AERIAL APPLICATION SETBACK DISTANCES FOR WOODLOTS AND FORESTS:

Occupied habitations	155 m
Surface water	60 m
Property boundaries	50 m
Municipal water sources	3.2 km from point of extraction

AERIAL APPLICATION SETBACK DISTANCES FOR AGRICULTURAL AREAS:

	HERBICIDES	INSECTICIDES
Occupied habitations	50 m	155 m
Surface water	65 m	100 m
Property boundaries	50 m	50 m

However, these are minimum distances and since all aerial application is by permit, additional distances can be imposed for very sensitivie situations and/or very highly toxic or very highly persistent pesticides.

NOVA SCOTIA:
Currently, no agricultural aerial application of pesticides is being carried out in Nova Scotia.

AERIAL APPLICATION SETBACK DISTANCES FOR WOODLOTS AND FORESTS:

Occupied habitations	100 m
Potable water sources	100 m
Farm crops	100 m
Watercourses	30 m

However, these are minimum distances and since all aerial application is by permit, additional distances can be imposed for very sensitive situations and/or very highly toxic or very highly persistent pesticides.

PRINCE EDWARD ISLAND:

AERIAL APPLICATION SETBACK DISTANCES

Occupied habitations	100 m
Open bodies of water	25 m

NEWFOUNDLAND
Not available at time of writing.

9.3 Specific Excerpts from FIFRA

FIFRA was enacted by U.S. Public Law 92-516, Oct. 21, 1972; Stat 975, 7 U.S.C. 136 et seq. and has been amended several times since. We give a number of informative excerpts below, from a copy dated 12-8-95 (Copyright © by the Bureau of National Affairs, Inc.).

FIFRA SECTION HEADINGS

Section 1. Short title and table of contents
Section 2. Definitions
Section 3. Registration of pesticides
Section 4. Reregistration of pesticides
Section 5. Experimental use permits
Section 6. Administrative review; suspension
Section 7. Registration of establishments
Section 8. Records
Section 9. Inspection of establishments, etc.
Section 10. Protection of trade secrets and other information
Section 11. Use of restricted use pesticides: applicators
Section 12. Unlawful acts

Section 13. Stop sale, use, removal, and seizure
Section 14. Penalties
Section 15. Indemnities
Section 16. Administrative procedure; judicial review
Section 17. Imports and exports
Section 18. Exemption of Federal and State agencies
Section 19. Storage, disposal, transportation, and recall
Section 20 Research and monitoring
Section 21. Solicitation of comments: notice of public hearings
Section 22. Delegation and cooperation
Section 23. State cooperation, aid, and training
Section 24. Authority of states
Section 25. Authority of administrator
Section 26. State primary enforcement responsibility
Section 27. Failure of the State to assure enforcement of State pesticide use regulations
Section 28. Identification of pests; cooperation with the Department of Agriculture's program
Section 29. Annual report
Section 30. Severability
Section 31. Authorization for appropriations

The act is administered by the Administrator of the Environmental Protection Agency. Distribution or sale of any pesticide that is not registered under FIFRA is prohibited (some exemptions are possible). The registration process involves an application to the Administrator including comprehensive data on the product. There are a large number of amendments to this section of the act dealing with the need and procedures for providing data regarding the nature and use of the pesticide.

Experimental use permits are allowed if the Administrator deems it necessary for the registration process. The Administrator may require that experimental studies be done to detect whether or not experimental use may cause adverse environmental effects.

Trade secrets are protected. However, information regarding the effects of the pesticide on an organism or on the environment are made available to the public. This proviso is qualified with amendments relating to disclosure of certain proprietary information relating to manufacturing processes, etc.

Applicator certification is the responsibility of the Administrator. Any State can submit a plan for certification to the Administrator for approval. In

States where applicator approval plans are not in place, the Administrator will conduct a program for certification of applicators in consultation with the Governor of the State.

The Administrator is empowered to execute a "stop sale, use, or removal" order with respect to any pesticide or device which is found on the basis of inspection (with respect to its nature or intended use) to be in violation of the provisions of the Act.

Information as to the storage, disposal, and transportation may be requested by the Administrator to support the registration of a pesticide. The Administrator may issue regulations as to the disposal of a pesticide in the case of cancellation of its registration.

9.4 The Training of Pesticide Applicators and the Inspection of Application Equipment

Operator training and application equipment inspection have been mentioned in sections **9.2** and **9.3**. With the kind permission of the American Society of Agricultural Engineers, we provide here a copy of a summary of a survey of US and European programs for training applicators and inspecting application equipment [Ozkan (1996)]:

a) Pesticide applicator education in the US and in Western Europe

Proper operation and calibration of pesticide application equipment should be an important part of the total knowledge that applicators are expected to have in order to apply pesticides accurately. The majority of underdeveloped and developing nations do not have the resources to implement educational programs to train pesticide applicators. Although generally the pesticide applicator education programs in the USA are the envy of many nations around the world, the part of the packaged program associated with selection, operation, and calibration of application equipment receives inadequate attention.

The procedure for teaching topics related to application equipment varies from state to state in the U.S., and from country to country around the world. In the US, every state training program must meet the minimum established in FIFRA requirements, but many states have their own state pesticide laws that are more stringent than the federal law. While some states address the topic adequately, most others allocate only 30 minutes to cover the important topic of application equipment. Such training processes and the educational activities associated with it, have not been very effective. They have not generated enough interest among applicators to calibrate their sprayers. In many cases, the large size of the audience attending these lecture-oriented teaching activities has not allowed audience participation and interaction with the instructor. In some cases, the subject matter is covered by using slide sets and videotapes with no instructor available to answer questions.

Typically, in the U.S., the applicators must pass a written exam to get a certificate to apply restricted use pesticides. For example, in Ohio, commercial applicators take an exam that contains 80 questions from the core material, and 25 to 35 questions from a specific category chosen by the applicant. Out of the approximately 110 questions, no more than 10 are directly associated with application equipment. Most of these questions, such as mixing chemicals, drift, nozzle selection, properly cleaning a nozzle, reasons for calibration, proper cleaning of a sprayer etc., can be answered by simply using common sense. Answering 65% of the total 110 questions will be sufficient for someone to pass the exam. Since the proportion of equipment and application related questions in the exam is low compared to the total number of questions, it is possible for someone (to) answer all of the equipment related questions wrong, yet score high enough to pass the exam and become a certified pesticide applicator.

On the other hand, there are some European countries (that) require an examination that includes a direct, one-on-one testing with spray equipment before a certified pesticide applicator rating is issued. Prior to this hands-on type exam, the applicator is required to go through a training school that is set up specifically for teaching the applicator the basic principles of sprayer operation and calibration, as well as proper handling of pesticides. These schools typically last 1 to 4 days depending on the experience of the applicator.

Another point of concern with the current applicator licensing process in the U.S. is that only the supervisor (is) required to hold a valid applicator license. Under his/her supervision, personnel who are not licensed may handle the pesticides and operate the spray equipment. The assumption is that the supervisor will provide adequate training to his/her employees before spraying begins. If supervision is close enough, this policy is adequate and can be practiced successfully. Unfortunately, there is no way of knowing whether or not supervisor training is practiced and adequate. In many European countries everyone who is actively engaged in applying pesticides has to obtain a valid applicator license.

The process of license renewal is different in the U.S. and in Western European countries. In most States in the U.S., renewal of a license is required every three years. The applicant must either take the test and pass, or attend educational programs for a required number of hours. During these structured educational programs the applicator is provided with the latest information on various aspects of pest control. Until recently, recertification of an operator was not necessary. Once a license was obtained, it was valid indefinitely. Although some European countries still do not require recertification, others have adopted the concept of recertification and continuing education that comes with the recertification process.

b) Inspection and certification of application equipment

All countries in this survey (Belgium, the Czech Republic, France, Germany, the Netherlands, Norway, Spain, Sweden, and the United Kingdom) have been conducting equipment inspection programs on a voluntary basis for several decades. The low level of participation by pesticide applicators in the voluntary inspection programs and increasing public concerns about the misuse of pesticides have forced regulators in some countries to implement much stricter rules governing the condition of pesticide application equipment. Germany and Belgium have made equipment inspection mandatory since January 1993 and September 1995, respectively. If a sprayer passes the mandatory inspection, a decal, representing the seal of approval by the regulatory agency of the government is placed on the sprayer. These inspections typically include (the) following items of concern: general condition of the sprayer (fittings, hoses, leakage, boom stability, pumps, nozzles, etc.), stability of pressure at the nozzle, condition of the pressure gage, pressure variations among the boom sections, change in pressure at the nozzle when any section of the boom is shut off, flow rates of all nozzles, accuracy of sprayer monitors and control units.

Within one month after the start of the mandatory inspection program in Belgium, 1566 sprayers were inspected. About 87% of these sprayers passed the inspection. The major reasons for rejection of sprayers included faulty pressure gages (43%) and nozzles (41%) with the wrong flow rate, angle or type. The goal of the Belgian government is to have all crop sprayers (about 28000) inspected before December 1998. In Belgium, the cost of the inspection depends on the length of the sprayer boom; costs range from 2500 Belgian Francs (US$82) for sprayers with booms less than 12 m long to 4000 BEF (US$132) for sprayers with booms greater than 22 m.

Currently, only boom-type field sprayers have been subjected to mandatory inspection. However, plans are underway in both Germany and Belgium to expand this program to include high-pressure orchard sprayers.

References

Irving, H.J. (1993): Aircraft Use and Regulations for Pesticide Application in Canada, available from the author, SERG Coordinator, 79 Silverwood Terrace, Fredericton, N.B., Can.,E3C1K2.

Mickle, R.E. (1996): A generic approach to setting buffer zones in Canada, in Biorational Pest Control Agents, Formulation and Delivery, F.R. Hall and J.W. Barry eds., ACS Symposium Series, pp 27-40.

Nicholson, I. (1994): The Canadian interdepartmental task force on pesticide drift, in Overview of Spray Drift and Deposit, Capco Technical Session, W. Sexsmith ed., Natural Resources Canada, Canadian Forest Service, Science and Sustainable Development, Hull, Quebec, pp 3-6.

Ozkan, H.E. (1996): Procedures for licensing pesticide applicators and certifying application equipment in Western Europe, paper #961083, 1996 ASAE Annual International Meeting, Phoenix, AZ, (July 1996), available at ASAE, 2950 Niles Rd. St. Joseph, MI, USA

Richardson, B. (1996): New Zealand Forest Research Institute, Plant Protection Chemistry, personal communication, July 1996.

APPENDIX I

SPRAY EFFICACY RESEARCH GROUP (SERG)

SERG is an association of forest management and research agencies interested in the application of forest and industrial management pest control products. The principal objective is the improvement of the technology associated with the use of pest control products, applied aerially, or from the ground. SERG coordinates research into 1) the efficacy of conventional and alternative products (virus, nematodes, biological herbicides, etc.), 2) application technology, 3) deposit and drift of applied products, and 3) improved operations such as GPS navigation. Environmental quality and technology transfer are key motives.

SERG COORDINATOR

Mr. H.J. Irving, 79 Silverwood Terrace, Fredericton, N.B., Can., E3C1K2; Tel. 506-452-1537, Fax 506-452-2962.

MEMBERS OF THE STEERING COMMITTEE, 1996

Churcher, Mr. J. J. (Joe), 70 Foster Drive, Suite 400, Ministry of Natural Resources,Sault Ste. Marie, ON,P6A 6J5.

Davies, Mr. David C., Forest Protection Limited, 2240 Lincoln Road, Fredericton, NB, E3B 7E6.

Dugal, Mr. Jacques, Société de protection des forêts contre les insectes et maladies, 1400 St. Jean Baptiste, Suite 200, Québec, QC, G2E 5B7.

Howse, Dr. Gordon, Canadian Forest Service - Ontario Region, Great Lakes Forestry Centre, 1219 Queen Street East, P.O. Box 490, Sault Ste. Marie, ON, P6A 5M7.

Kettela, Mr. Edward G., Canadian Forest Service - Maritimes, Natural Resources Canada, P.O. Box 4000, Fredericton, NB, E3B 5P7

Matwee, Mr. Larry, Operations Manager, Forestry Branch, Province of Manitoba, Box 27, Group 8, R.R.#2, Dugald, MB, R0E 2K0.

McCooeye, Ms. M.M., Institute for Environmental Research & Technology, National Research Council Canada, Montreal Road, M-12, Room G-17, Ottawa, ON, K1A 0R6.

Mickle, Dr. R.E., Air Quality Research Branch, Atmospheric Environment Service, 4905 Dufferin Street, Downsview, ON, N0B 1E0.

Moody, Dr. B.H., Natural Resources Canada, 351 St. Joseph Boulevard, Hull, Quebec, K1A 1G5.

Régnière, Dr. Jacques, Entomologist, Population Dynamics, Forestry Canada, Quebec Region, Laurentian Forestry Centre, 1055 du P.E.P.S., P.O. Box 3800 Sainte-Foy, QC, G1V 4C7.

Sexsmith, Ms. Wendy, Director, Alternatives Office, Pest Management, Regulatory Agency, 59 Camelot Drive, Nepean, Ontario, K1A 0Y9.

Sterner, Dr. Tom, Forestry Canada - Maritimes, Natural Resources Canada, P.O. Box 4000, Fredericton, NB, E3B 5P7.

Volney, Dr. Jan, Natural Resources Canada, Canadian Forestry Service, Northern Forestry Centre, 5320 - 122nd St., Edmonton, AB, T6H 3S5.

Westwood, Dr. Richard, Manitoba Natural Resources, Forestry Branch, Forest Protection, 300 530 Kenaston Blvd., Winnipeg, MB, R3N 1Z4.

APPENDIX II

ATOMIZER CHARACTERIZATION DATA

CONTENTS:

NOTES:

A) All tables are based on total plume surveys (see Chapter 4, sections 4.4.1 and 4.4.2.

B) Malvern spectrometer results show a "mass balance closure" defined as measured plume droplet volumetric flow rate divided by flow rate supplied to the atomizer, expressed as %. Deviations from 100% are caused by high obscurations (low values), and errors in the control of wind tunnel velocity and atomizer flow rate, both random and of a maximum magnitude of about 2% each. Typically, 12 positions (separated by 5 cms) of the Malvern laser beam are required to encompass the spray plumes. Our experience is that when 50% of the "shots" give obscuration between 0.10 and 0.20 and the rest, lower values, mass balance closure is 80% or better. Work continues on a statistical method of correcting results with lower mass balance closures.

C) In Malvern results tables, the number fractions for each bin can be found by dividing the value of Noi by the total Noi appearing at the bottom of the column. Cumulative number fractions, etc. follow from these values.

Nozzle: AU4000 Focal Length (mm): 600 Surface tension 40 dynes/cm
Flowrate (l/min): 12.6 Sweeps: 2500
Wind Speed (m/s): 45 Malvern Filename: d14oct94
Rotational Rate (RPM): 3500 Fluid: 9% (vol) isopropanol in water

Lower Diameter (microns)	Upper Diameter (microns)	No. of Drops Noi (No.)	Volumetric Rate Fi (m^3/s)	Droplet Number Rate Ni (No./s)		Ni/DD	CVF	Calc. Droplet Velocity (m/s)
3.0	11.6	8.6	1.539e+05	4.031e-07	1.979e+09	2.301e+08	0.002	55.0
11.6	13.4	1.8	2.827e+04	3.718e-07	3.635e+08	2.020e+08	0.003	55.0
13.4	15.6	2.2	1.923e+04	3.948e-07	2.473e+08	1.124e+08	0.005	55.0
15.6	18.1	2.5	1.376e+04	4.432e-07	1.769e+08	7.077e+07	0.007	55.0
18.1	20.9	2.8	1.437e+04	7.165e-07	1.845e+08	6.591e+07	0.010	54.9
20.9	24.3	3.4	1.607e+04	1.245e-06	2.060e+08	6.058e+07	0.015	54.8
24.3	28.1	3.8	1.361e+04	1.640e-06	1.741e+08	4.582e+07	0.022	54.7
28.1	32.6	4.5	9.409e+03	1.752e-06	1.197e+08	2.660e+07	0.029	54.4
32.6	37.8	5.2	6.698e+03	1.935e-06	8.473e+07	1.629e+07	0.037	54.1
37.8	43.8	6.0	6.049e+03	2.701e-06	7.596e+07	1.266e+07	0.048	53.7
43.8	50.3	6.5	6.069e+03	4.131e-06	7.578e+07	1.168e+07	0.065	53.4
50.3	61.0	10.7	5.079e+03	5.676e-06	6.295e+07	5.889e+06	0.089	53.0
61.0	73.3	12.3	3.396e+03	6.599e-06	4.168e+07	3.396e+06	0.116	52.5
73.3	87.6	14.4	2.154e+03	7.154e-06	2.625e+07	1.825e+06	0.146	52.1
87.6	104.3	16.6	1.399e+03	7.807e-06	1.688e+07	1.015e+06	0.178	51.6
104.3	123.1	18.9	1.012e+03	9.305e-06	1.209e+07	6.407e+05	0.217	51.1
123.1	145.6	22.4	8.764e+02	1.314e-05	1.035e+07	4.613e+05	0.271	50.5
145.6	172.0	26.4	8.225e+02	2.011e-05	9.597e+06	3.636e+05	0.354	49.9
172.0	201.0	29.0	6.592e+02	2.585e-05	7.615e+06	2.623e+05	0.461	49.4
201.0	236.6	35.6	5.061e+02	3.168e-05	5.775e+06	1.621e+05	0.593	48.8
236.6	276.2	39.6	3.142e+02	3.127e-05	3.542e+06	8.946e+04	0.722	48.2
276.2	323.7	47.5	1.779e+02	2.792e-05	1.976e+06	4.159e+04	0.838	47.5
323.7	377.8	54.1	7.527e+01	1.862e-05	8.237e+05	1.522e+04	0.915	46.8
377.8	439.9	62.0	2.068e+01	7.978e-06	2.230e+05	3.595e+03	0.948	46.1
439.9	512.4	72.6	8.788e+00	5.262e-06	9.309e+04	1.282e+03	0.970	45.3
512.4	596.9	84.5	4.438e+00	4.126e-06	4.618e+04	5.468e+02	0.987	44.5
596.9	694.6	97.7	1.749e+00	2.519e-06	1.787e+04	1.830e+02	0.997	43.7
694.6	808.0	113.5	2.232e-01	4.972e-07	2.239e+03	1.973e+01	0.999	42.9
808.0	938.7	130.6	2.653e-02	9.088e-08	2.605e+02	1.994e+00	1.000	42.0
938.7	1091.8	153.1	1.106e-02	5.823e-08	1.063e+02	6.942e-01	1.000	41.1
1091.8	1267.3	175.5	1.958e-03	1.586e-08	1.845e+01	1.051e-01	1.000	40.3
1267.3	1471.8	204.5	0.000e+00	0.000e+00	0.000e+00	0.000e+00	1.000	39.3
	Total			3.039e+05	2.414e-04	3.887e+09		

$D_{v0.1} = 66$ $D_{v0.5} = 212$ $D_{v0.9} = 367$ (microns) Mass balance closure 115%

Nozzle: AU4000
Flowrate (l/min): 12.6
Wind Speed (m/s): 45
Rotational Rate (RPM): 7900

Focal Length (mm): 600
Sweeps: 2500
Malvern Filename: d22oct94
Fluid: 9% (VOL) isopropanol in water

Surface tension 40 dynes/cm

Lower Diameter (microns)	Upper Diameter (microns)	DD	No. of Drops Noi (No.)	Volumetric Rate Fi (m^3/s)	Droplet Number Rate Ni (No./s)	Ni/DD	CVF	Calc. Droplet Velocity (m/s)
3.0	11.6	8.6	1.042e+06	2.729e-06	1.340e+10	1.558e+09	0.013	55.0
11.6	13.4	1.8	1.373e+05	1.806e-06	1.766e+09	9.811e+08	0.021	55.0
13.4	15.6	2.2	7.939e+04	1.630e-06	1.021e+09	4.641e+08	0.029	55.0
15.6	18.1	2.5	4.737e+04	1.526e-06	6.092e+08	2.437e+08	0.036	55.0
18.1	20.9	2.8	4.107e+04	2.047e-06	5.272e+08	1.883e+08	0.046	54.9
20.9	24.3	3.4	4.539e+04	3.516e-06	5.817e+08	1.711e+08	0.062	54.8
24.3	28.1	3.8	4.029e+04	4.853e-06	5.154e+08	1.356e+08	0.085	54.7
28.1	32.6	4.5	3.117e+04	5.804e-06	3.965e+08	8.811e+07	0.112	54.4
32.6	37.8	5.2	2.177e+04	6.288e-06	2.753e+08	5.295e+07	0.142	54.1
37.8	43.8	6.0	1.638e+04	7.313e-06	2.056e+08	3.427e+07	0.176	53.7
43.8	50.3	6.5	1.422e+04	9.680e-06	1.776e+08	2.736e+07	0.221	53.4
50.3	61.0	10.7	1.170e+04	1.307e-05	1.450e+08	1.356e+07	0.283	53.0
61.0	73.3	12.3	8.700e+03	1.691e-05	1.068e+08	8.702e+06	0.362	52.5
73.3	87.6	14.4	6.076e+03	2.018e-05	7.402e+07	5.146e+06	0.457	52.1
87.6	104.3	16.6	3.918e+03	2.186e-05	4.727e+07	2.843e+06	0.559	51.6
104.3	123.1	18.9	2.442e+03	2.246e-05	2.918e+07	1.546e+06	0.665	51.1
123.1	145.6	22.4	1.513e+03	2.269e-05	1.787e+07	7.966e+05	0.771	50.5
145.6	172.0	26.4	7.954e+02	1.945e-05	9.282e+06	3.517e+05	0.862	49.9
172.0	201.0	29.0	3.494e+02	1.370e-05	4.036e+06	1.390e+05	0.927	49.4
201.0	236.6	35.6	1.255e+02	7.854e-06	1.432e+06	4.019e+04	0.963	48.8
236.6	276.2	39.6	4.228e+01	4.207e-06	4.765e+05	1.204e+04	0.983	48.2
276.2	323.7	47.5	1.311e+01	2.057e-06	1.456e+05	3.064e+03	0.993	47.5
323.7	377.8	54.1	3.325e+00	8.224e-07	3.639e+04	6.726e+02	0.997	46.8
377.8	439.9	62.0	7.707e-01	2.973e-07	8.308e+03	1.340e+02	0.998	46.1
439.9	512.4	72.6	2.733e-01	1.636e-07	2.895e+03	3.988e+01	0.999	45.3
512.4	596.9	84.5	1.203e-01	1.118e-07	1.252e+03	1.482e+01	0.999	44.5
596.9	694.6	97.7	2.075e-02	2.989e-08	2.120e+02	2.171e+00	0.999	43.7
694.6	808.0	113.5	1.200e-02	2.673e-08	1.204e+02	1.061e+00	1.000	42.9
808.0	938.7	130.6	7.640e-03	2.617e-08	7.503e+01	5.743e-01	1.000	42.0
938.7	1091.8	153.1	3.891e-03	2.049e-08	3.739e+01	2.443e-01	1.000	41.1
1091.8	1267.3	175.5	1.861e-03	1.507e-08	1.753e+01	9.990e-02	1.000	40.3
1267.3	1471.8	204.5	1.585e-03	1.959e-08	1.456e+01	7.120e-02	1.000	39.3
Total			1.552e+06	2.132e-04	1.991e+10			

$D_{v0.1}$ = 31 $D_{v0.5}$ = 95 $D_{v0.9}$ = 189 (microns) Mass balance closure = 103%

Nozzle: AU4000
Flowrate (l/min): 12.6
Wind Speed (m/s): 45
Rotational Rate (RPM): 10400

Focal Length (mm): 600
Sweeps: 2500
Malvern Filename: d26oct94
Fluid: 9% isopropanol in water

Surface tension 40 dynes/cm

Lower Diameter (microns)	Upper Diameter (microns)	DD	No. of Drops Noi (No.)	Volumetric Rate Fi (m^3/s)	Droplet Number Rate Ni (No./s)	Ni/DD	CVF	Calc. Droplet Velocity (m/s)
3.0	11.6	8.6	1.154e+06	3.023e-06	1.484e+10	1.726e+09	0.036	55.0
11.6	13.4	1.8	8.587e+04	1.129e-06	1.104e+09	6.136e+08	0.049	55.0
13.4	15.6	2.2	3.905e+04	8.017e-07	5.022e+08	2.283e+08	0.058	55.0
15.6	18.1	2.5	1.031e+04	3.323e-07	1.326e+08	5.306e+07	0.062	55.0
18.1	20.9	2.8	1.001e+04	4.987e-07	1.285e+08	4.588e+07	0.068	54.9
20.9	24.3	3.4	2.681e+04	2.077e-06	3.436e+08	1.010e+08	0.092	54.8
24.3	28.1	3.8	2.867e+04	3.453e-06	3.667e+08	9.651e+07	0.133	54.7
28.1	32.6	4.5	1.926e+04	3.587e-06	2.451e+08	5.446e+07	0.175	54.4
32.6	37.8	5.2	1.142e+04	3.300e-06	1.445e+08	2.779e+07	0.214	54.1
37.8	43.8	6.0	8.204e+03	3.663e-06	1.030e+08	1.717e+07	0.257	53.7
43.8	50.3	6.5	8.748e+03	5.955e-06	1.092e+08	1.683e+07	0.327	53.4
50.3	61.0	10.7	7.472e+03	8.349e-06	9.260e+07	8.663e+06	0.425	53.0
61.0	73.3	12.3	5.095e+03	9.902e-06	6.255e+07	5.097e+06	0.542	52.5
73.3	87.6	14.4	3.336e+03	1.108e-05	4.065e+07	2.826e+06	0.672	52.1
87.6	104.3	16.6	1.721e+03	9.604e-06	2.076e+07	1.249e+06	0.785	51.6
104.3	123.1	18.9	8.111e+02	7.459e-06	9.692e+06	5.136e+05	0.873	51.1
123.1	145.6	22.4	3.601e+02	5.400e-06	4.252e+06	1.896e+05	0.936	50.5
145.6	172.0	26.4	1.244e+02	3.042e-06	1.451e+06	5.499e+04	0.972	49.9
172.0	201.0	29.0	3.492e+01	1.370e-06	4.034e+05	1.390e+04	0.988	49.4
201.0	236.6	35.6	1.008e+01	6.310e-07	1.150e+05	3.228e+03	0.996	48.8
236.6	276.2	39.6	1.990e+00	1.980e-07	2.243e+04	5.665e+02	0.998	48.2
276.2	323.7	47.5	5.167e-01	8.111e-08	5.739e+03	1.208e+02	0.999	47.5
323.7	377.8	54.1	1.875e-01	4.637e-08	2.052e+03	3.792e+01	1.000	46.8
377.8	439.9	62.0	1.056e-02	4.073e-09	1.138e+02	1.835e+00	1.000	46.1
439.9	512.4	72.6	3.699e-02	2.215e-08	3.919e+02	5.399e+00	1.000	45.3
512.4	596.9	84.5	0.000e+00	0.000e+00	0.000e+00	0.000e+00	1.000	44.5
596.9	694.6	97.7	0.000e+00	0.000e+00	0.000e+00	0.000e+00	1.000	43.7
694.6	808.0	113.5	0.000e+00	0.000e+00	0.000e+00	0.000e+00	1.000	42.9
808.0	938.7	130.6	0.000e+00	0.000e+00	0.000e+00	0.000e+00	1.000	42.0
938.7	1091.8	153.1	6.896e-04	3.631e-09	6.628e+00	4.330e-02	1.000	41.1
1091.8	1267.3	175.5	4.397e-04	3.561e-09	4.144e+00	2.361e-02	1.000	40.3
1267.3	1471.8	204.5	2.809e-04	3.472e-09	2.582e+00	1.262e-02	1.000	39.3
Total			1.421e+06	8.502e-05	1.825e+10			

$D_{v0.1} = 25$ $D_{v0.5} = 69$ $D_{v0.9} = 133$ Mass balance closure 81%

Nozzle: AU4000
Flowrate (l/min): 12.6
Wind Speed (m/s): 45
Rotational Rate (RPM): 10400

Focal Length (mm): 300
Sweeps: 2500
Malvern Filename: t26oct94
Fluid: 9% isopropanol in water

Surface tension 40 dynes/cm

Lower Diameter (microns)	Upper Diameter (microns)	DD	No. of Drops Noi (No.)	Volumetric Rate Fi (m^3/s)	Droplet Number Rate Ni (No./s)	Ni/DD	CVF	Calc. Droplet Velocity (m/s)
1.5	5.8	4.3	6.307e+04	2.065e-08	8.111e+08	1.886e+08	0.000	55.0
5.8	6.7	0.9	0.000e+00	0.000e+00	0.000e+00	0.000e+00	0.000	55.0
6.7	7.8	1.1	0.000e+00	0.000e+00	0.000e+00	0.000e+00	0.000	55.0
7.8	9.1	1.3	5.660e+03	2.279e-08	7.279e+07	5.823e+07	0.000	55.0
9.1	10.4	1.4	1.104e+04	6.838e-08	1.420e+08	1.052e+08	0.001	55.0
10.4	12.1	1.7	5.987e+04	5.740e-07	7.699e+08	4.529e+08	0.008	55.0
12.1	14.1	2.0	1.974e+04	2.988e-07	2.538e+08	1.269e+08	0.011	55.0
14.1	16.3	2.2	4.085e+04	9.660e-07	5.254e+08	2.388e+08	0.022	55.0
16.3	18.9	2.6	1.013e+05	3.719e-06	1.303e+09	5.011e+08	0.064	55.0
18.9	21.9	3.0	7.165e+04	4.088e-06	9.198e+08	3.066e+08	0.110	54.9
21.9	25.4	3.5	2.492e+04	2.211e-06	3.193e+08	9.123e+07	0.135	54.8
25.4	29.5	4.1	8.252e+03	1.143e-06	1.055e+08	2.574e+07	0.147	54.7
29.5	34.1	4.6	9.236e+03	1.978e-06	1.175e+08	2.554e+07	0.170	54.4
34.1	39.5	5.4	1.006e+04	3.320e-06	1.272e+08	2.356e+07	0.207	54.1
39.5	45.8	6.3	1.503e+04	7.665e-06	1.887e+08	2.995e+07	0.293	53.7
45.8	53.0	7.2	9.576e+03	7.548e-06	1.196e+08	1.661e+07	0.378	53.4
53.0	61.5	8.5	6.370e+03	7.757e-06	7.895e+07	9.288e+06	0.465	53.0
61.5	71.5	10.0	5.288e+03	1.002e-05	6.505e+07	6.505e+06	0.578	52.6
71.5	82.5	11.0	3.674e+03	1.072e-05	4.484e+07	4.077e+06	0.699	52.2
82.5	96.6	14.1	2.193e+03	9.969e-06	2.651e+07	1.881e+06	0.811	51.7
96.6	112.9	16.3	9.748e+02	7.032e-06	1.169e+07	7.191e+05	0.890	51.3
112.9	132.4	19.5	3.889e+02	4.458e-06	4.619e+06	2.367e+05	0.940	50.8
132.4	155.1	22.8	1.422e+02	2.606e-06	1.676e+06	7.362e+04	0.969	50.4
155.1	180.1	24.9	4.766e+01	1.373e-06	5.572e+05	2.235e+04	0.985	50.0
180.1	210.4	30.4	1.647e+01	7.427e-07	1.906e+05	6.280e+03	0.993	49.5
210.4	245.1	34.7	4.477e+00	3.167e-07	5.119e+04	1.476e+03	0.997	48.9
245.1	285.2	40.1	1.437e+00	1.584e-07	1.623e+04	4.047e+02	0.998	48.3
285.2	331.8	46.6	3.195e-01	5.480e-08	3.564e+03	7.646e+01	0.999	47.7
331.8	384.9	53.1	2.126e-01	5.633e-08	2.337e+03	4.400e+01	1.000	47.0
384.9	447.8	62.9	1.866e-03	7.636e-10	2.020e+01	3.213e-01	1.000	46.3
447.8	520.4	72.6	1.187e-03	7.504e-10	1.263e+01	1.739e-01	1.000	45.5
520.4	603.9	83.5	2.595e-02	2.517e-08	2.706e+02	3.242e+00	1.000	44.6
Total			4.693e+05	8.891e-05	6.010e+09			

$D_{v0.1} = 21$ $D_{v0.5} = 65$ $D_{v0.9} = 117$ (microns) Mass balance closure 85%

Atomizer: AU4000
Wind Speed (m/s): 56
Flowrate (l/min): 75
Malvern Filename: d18dec94

Rotational Rate (RPM): 2700
Focal Length (mm): 800
Sweeps: 1000
Fluid: 29% isopropanol in water

Surface tension 31 dynes/cm

Lower Diameter (microns)	Upper Diameter (microns)	DD	No. of Drops Noi (No.)	Volumetric Rate Fi (m^3/s)	Droplet Number Rate Ni (No./s)	Ni/DD	CVF	Calc. Droplet Velocity (m/s)
13.5	25.0	11.5	6.533e+04	3.195e-06	8.556e+08	7.440e+07	0.008	56.0
25.0	27.4	2.4	1.233e+04	1.520e-06	1.614e+08	6.725e+07	0.012	56.0
27.4	30.3	2.9	8.351e+03	1.375e-06	1.094e+08	3.771e+07	0.015	56.0
30.3	33.6	3.3	3.620e+03	8.095e-07	4.741e+07	1.437e+07	0.017	56.0
33.6	37.4	3.8	2.316e+03	7.104e-07	3.033e+07	7.981e+06	0.019	56.0
37.4	41.9	4.5	3.948e+03	1.687e-06	5.170e+07	1.149e+07	0.023	56.0
41.9	47.0	5.1	5.865e+03	3.532e-06	7.680e+07	1.506e+07	0.032	56.0
47.0	53.0	6.0	7.125e+03	6.107e-06	9.331e+07	1.555e+07	0.048	56.0
53.0	59.9	6.9	5.534e+03	6.826e-06	7.247e+07	1.050e+07	0.065	56.0
59.9	67.9	8.0	3.235e+03	5.788e-06	4.237e+07	5.296e+06	0.079	56.0
67.9	77.2	9.3	2.050e+03	5.366e-06	2.684e+07	2.886e+06	0.093	56.0
77.2	88.0	10.8	1.846e+03	7.134e-06	2.418e+07	2.239e+06	0.111	56.0
88.0	100.4	12.4	1.989e+03	1.140e-05	2.605e+07	2.101e+06	0.140	56.0
100.4	114.5	14.1	1.846e+03	1.570e-05	2.417e+07	1.714e+06	0.179	56.0
114.5	131.5	17.0	1.435e+03	1.831e-05	1.879e+07	1.105e+06	0.225	56.0
131.5	150.5	19.0	9.988e+02	1.920e-05	1.308e+07	6.884e+05	0.273	56.0
150.5	173.5	23.0	8.107e+02	2.363e-05	1.062e+07	4.616e+05	0.333	56.0
173.5	200.5	27.0	7.859e+02	3.524e-05	1.029e+07	3.812e+05	0.422	56.0
200.5	229.5	29.0	6.380e+02	4.347e-05	8.354e+06	2.881e+05	0.531	56.0
229.5	265.5	36.0	4.919e+02	5.113e-05	6.441e+06	1.789e+05	0.660	56.0
265.5	305.5	40.0	3.019e+02	4.818e-05	3.954e+06	9.884e+04	0.781	56.0
305.5	353.5	48.0	1.760e+02	4.317e-05	2.305e+06	4.802e+04	0.890	56.0
353.5	408.5	55.0	7.344e+01	2.785e-05	9.616e+05	1.748e+04	0.960	56.0
408.5	470.5	62.0	1.406e+01	8.181e-06	1.840e+05	2.969e+03	0.980	56.0
470.5	544.5	74.0	4.383e+00	3.928e-06	5.739e+04	7.755e+02	0.990	56.0
544.5	629.5	85.0	2.132e+00	2.957e-06	2.792e+04	3.285e+02	0.998	56.0
629.5	728.5	99.0	4.595e-01	9.862e-07	6.017e+03	6.078e+01	1.000	56.0
728.5	842.5	114.0	0.000e+00	0.000e+00	0.000e+00	0.000e+00	1.000	56.0
842.5	974.5	132.0	0.000e+00	0.000e+00	0.000e+00	0.000e+00	1.000	56.0
974.5	1129.5	155.0	0.000e+00	0.000e+00	0.000e+00	0.000e+00	1.000	56.0
1129.5	1306.5	177.0	0.000e+00	0.000e+00	0.000e+00	0.000e+00	1.000	56.0
1306.5	1513.5	207.0	0.000e+00	0.000e+00	0.000e+00	0.000e+00	1.000	56.0
			1.311e+05	3.974e-04	1.717e+09			

$D_{v0.1}$ = 81 $D_{v0.5}$ = 221 $D_{v0.9}$ = 361 (Microns) Mass balance closure 64%

Atomizer: AU4000
Wind Speed (m/s): 45
Flowrate (l/min): 72
Malvern Filename: d17dec94

Rotational Rate (RPM): 2600
Focal Length (mm): 800
Sweeps: 1000
Fluid: 295 isopropanol in water

Surface tension 31 dynws/cm

Lower Diameter (microns)	Upper Diameter (microns)	DD	No. of Drops Noi (No.)	Volumetric Rate Fi (m^3/s)	Droplet Number Rate Ni (No./s)	Ni/DD	CVF	Calc. Droplet Velocity (m/s)
13.5	25.0	11.5	1.144e+03	4.498e-08	1.204e+07	1.047e+06	0.000	45.0
25.0	27.4	2.4	4.539e+02	4.498e-08	4.776e+06	1.990e+06	0.000	45.0
27.4	30.3	2.9	1.430e+03	1.892e-07	1.505e+07	5.190e+06	0.001	45.0
30.3	33.6	3.3	2.054e+03	3.691e-07	2.162e+07	6.550e+06	0.002	45.0
33.6	37.4	3.8	3.339e+03	8.230e-07	3.513e+07	9.245e+06	0.003	45.0
37.4	41.9	4.5	2.947e+03	1.012e-06	3.101e+07	6.892e+06	0.006	45.0
41.9	47.0	5.1	2.526e+03	1.222e-06	2.658e+07	5.213e+06	0.009	45.0
47.0	53.0	6.0	1.741e+03	1.199e-06	1.832e+07	3.053e+06	0.011	45.0
53.0	59.9	6.9	1.394e+03	1.381e-06	1.467e+07	2.126e+06	0.015	45.0
59.9	67.9	8.0	1.318e+03	1.895e-06	1.387e+07	1.734e+06	0.019	45.0
67.9	77.2	9.3	1.798e+03	3.783e-06	1.892e+07	2.034e+06	0.028	45.0
77.2	88.0	10.8	1.636e+03	5.080e-06	1.722e+07	1.594e+06	0.040	45.0
88.0	100.4	12.4	1.134e+03	5.224e-06	1.194e+07	9.625e+05	0.052	45.0
100.4	114.5	14.1	6.707e+02	4.584e-06	7.058e+06	5.006e+05	0.063	45.0
114.5	131.5	17.0	5.121e+02	5.250e-06	5.388e+06	3.170e+05	0.075	45.0
131.5	150.5	19.0	6.284e+02	9.705e-06	6.612e+06	3.480e+05	0.098	45.0
150.5	173.5	23.0	6.769e+02	1.586e-05	7.122e+06	3.097e+05	0.135	45.0
173.5	200.5	27.0	4.974e+02	1.792e-05	5.234e+06	1.938e+05	0.177	45.0
200.5	229.5	29.0	3.345e+02	1.832e-05	3.520e+06	1.214e+05	0.220	45.0
229.5	265.5	36.0	2.928e+02	2.446e-05	3.081e+06	8.559e+04	0.277	45.0
265.5	305.5	40.0	2.293e+02	2.940e-05	2.413e+06	6.032e+04	0.346	45.0
305.5	353.5	48.0	2.413e+02	4.755e-05	2.539e+06	5.289e+04	0.457	45.0
353.5	408.5	55.0	2.100e+02	6.399e-05	2.210e+06	4.017e+04	0.607	45.0
408.5	470.5	62.0	1.332e+02	6.229e-05	1.401e+06	2.260e+04	0.753	45.0
470.5	544.5	74.0	7.051e+01	5.078e-05	7.420e+05	1.003e+04	0.872	45.0
544.5	629.5	85.0	3.071e+01	3.422e-05	3.231e+05	3.801e+03	0.952	45.0
629.5	728.5	99.0	7.474e+00	1.289e-05	7.864e+04	7.944e+02	0.983	45.0
728.5	842.5	114.0	1.153e+00	3.078e-06	1.213e+04	1.064e+02	0.990	45.0
842.5	974.5	132.0	4.654e-01	1.923e-06	4.898e+03	3.710e+01	0.994	45.0
974.5	1129.5	155.0	1.907e-01	1.223e-06	2.007e+03	1.295e+01	0.997	45.0
1129.5	1306.5	177.0	8.230e-02	8.193e-07	8.660e+02	4.893e+00	0.999	45.0
1306.5	1513.5	207.0	2.629e-02	4.060e-07	2.766e+02	1.336e+00	1.000	45.0
			2.745e+04	4.269e-04	2.889e+08			

$D_{v0.1}$ = 152 $D_{v0.5}$ = 369 $D_{v0.9}$ = 574 (microns) Mass balance closure 71%

Atomizer: AU4000
Wind Speed (m/s): 45
Flowrate (l/min): 38
Malvern Filename: d07dec94

Rotational Rate (RPM): 5500
Focal Length (mm): 600
Sweeps: 1000
Fluid: 29% isopropanol in water

Surface tension 31 dynes/cm

Lower Diameter (microns)	Upper Diameter (microns)	DD	No. of Drops Noi (No.)	Volumetric Rate Fi (m^3/s)	Droplet Number Rate Ni (No./s)	Ni/DD	CVF	Calc. Droplet Velocity (m/s)
3.0	11.6	8.6	4.911e+05	1.286e-06	6.316e+09	7.344e+08	0.006	55.0
11.6	13.4	1.8	6.187e+04	8.138e-07	7.958e+08	4.421e+08	0.010	55.0
13.4	15.6	2.2	4.609e+04	9.462e-07	5.928e+08	2.694e+08	0.014	55.0
15.6	18.1	2.5	2.797e+04	9.012e-07	3.598e+08	1.439e+08	0.018	55.0
18.1	20.9	2.8	2.734e+04	1.363e-06	3.510e+08	1.254e+08	0.025	54.9
20.9	24.3	3.4	2.615e+04	2.025e-06	3.351e+08	9.856e+07	0.034	54.8
24.3	28.1	3.8	2.113e+04	2.546e-06	2.703e+08	7.114e+07	0.046	54.7
28.1	32.6	4.5	1.533e+04	2.854e-06	1.950e+08	4.333e+07	0.059	54.4
32.6	37.8	5.2	1.148e+04	3.315e-06	1.452e+08	2.792e+07	0.075	54.1
37.8	43.8	6.0	9.058e+03	4.045e-06	1.137e+08	1.896e+07	0.093	53.7
43.8	50.3	6.5	8.454e+03	5.757e-06	1.056e+08	1.624e+07	0.120	53.4
50.3	61.0	10.7	6.549e+03	7.324e-06	8.116e+07	7.585e+06	0.154	53.0
61.0	73.3	12.3	4.262e+03	8.294e-06	5.232e+07	4.253e+06	0.193	52.5
73.3	87.6	14.3	2.720e+03	9.036e-06	3.314e+07	2.318e+06	0.235	52.1
87.6	104.3	16.7	1.725e+03	9.626e-06	2.081e+07	1.246e+06	0.280	51.6
104.3	123.1	18.8	1.372e+03	1.262e-05	1.640e+07	8.723e+05	0.338	51.1
123.1	145.6	22.5	1.135e+03	1.702e-05	1.340e+07	5.956e+05	0.418	50.5
145.6	172.0	26.4	9.188e+02	2.248e-05	1.072e+07	4.061e+05	0.522	49.9
172.0	201.0	29.0	6.739e+02	2.644e-05	7.784e+06	2.684e+05	0.645	49.4
201.0	236.6	35.6	4.206e+02	2.632e-05	4.799e+06	1.348e+05	0.767	48.8
236.6	276.2	39.6	2.004e+02	1.993e-05	2.259e+06	5.704e+04	0.860	48.2
276.2	323.7	47.5	9.009e+01	1.414e-05	1.001e+06	2.107e+04	0.926	47.5
323.7	377.8	54.1	3.515e+01	8.690e-06	3.846e+05	7.110e+03	0.966	46.8
377.8	439.9	62.1	9.532e+00	3.677e-06	1.028e+05	1.655e+03	0.984	46.1
439.9	512.4	72.5	3.095e+00	1.853e-06	3.278e+04	4.522e+02	0.992	45.3
512.4	596.9	84.5	1.053e+00	9.786e-07	1.095e+04	1.296e+02	0.997	44.5
596.9	694.6	97.7	3.646e-01	5.253e-07	3.726e+03	3.814e+01	0.999	43.7
694.6	808.0	113.4	7.454e-02	1.660e-07	7.478e+02	6.594e+00	1.000	42.9
808.0	938.7	130.7	7.120e-03	2.439e-08	6.992e+01	5.350e-01	1.000	42.0
938.7	1091.8	153.1	0.000e+00	0.000e+00	0.000e+00	0.000e+00	1.000	41.1
1091.8	1267.3	175.5	0.000e+00	0.000e+00	0.000e+00	0.000e+00	1.000	40.3
1267.3	1471.8	204.5	0.000e+00	0.000e+00	0.000e+00	0.000e+00	1.000	39.3
			7.661e+05	2.150e-04	9.825e+09			

$D_{v0.1}$ =45 $D_{v0.5}$ = 166 $D_{v0.9}$ = 305 (microns) Mass balance closure 68%

Atomizer: AU4000
Wind Speed (m/s): 45
Flowrate (l/min): 38
Malvern Filename: d06dec94

Rotational Rate (RPM): 7000
Focal Length (mm): 600
Sweeps: 1000
Fluid: 29% isopropanol in water

Surface tension 31 dynes/cm

Lower Diameter (microns)	Upper Diameter (microns)	DD	No. of Drops Noi (No.)	Volumetric Rate Fi (m^3/s)	Droplet Number Rate Ni (No./s)	Ni/DD	CVF	Calc. Droplet Velocity (m/s)
3.0	11.6	8.6	1.069e+06	2.800e-06	1.375e+10	1.599e+09	0.014	55.0
11.6	13.4	1.8	1.261e+05	1.659e-06	1.622e+09	9.010e+08	0.022	55.0
13.4	15.6	2.2	8.000e+04	1.642e-06	1.029e+09	4.677e+08	0.030	55.0
15.6	18.1	2.5	4.979e+04	1.604e-06	6.403e+08	2.561e+08	0.038	55.0
18.1	20.9	2.8	4.031e+04	2.009e-06	5.174e+08	1.848e+08	0.048	54.9
20.9	24.3	3.4	4.238e+04	3.283e-06	5.431e+08	1.597e+08	0.065	54.8
24.3	28.1	3.8	3.593e+04	4.328e-06	4.596e+08	1.209e+08	0.086	54.7
28.1	32.6	4.5	2.686e+04	5.001e-06	3.416e+08	7.592e+07	0.111	54.4
32.6	37.8	5.2	1.977e+04	5.711e-06	2.501e+08	4.810e+07	0.140	54.1
37.8	43.8	6.0	1.534e+04	6.852e-06	1.927e+08	3.211e+07	0.174	53.7
43.8	50.3	6.5	1.331e+04	9.062e-06	1.662e+08	2.556e+07	0.219	53.4
50.3	61.0	10.7	9.895e+03	1.107e-05	1.226e+08	1.146e+07	0.274	53.0
61.0	73.3	12.3	6.105e+03	1.188e-05	7.494e+07	6.093e+06	0.334	52.5
73.3	87.6	14.3	3.909e+03	1.298e-05	4.762e+07	3.330e+06	0.398	52.1
87.6	104.3	16.7	2.556e+03	1.426e-05	3.084e+07	1.846e+06	0.469	51.6
104.3	123.1	18.8	1.849e+03	1.700e-05	2.209e+07	1.175e+06	0.554	51.1
123.1	145.6	22.5	1.421e+03	2.131e-05	1.678e+07	7.459e+05	0.661	50.5
145.6	172.0	26.4	8.969e+02	2.194e-05	1.047e+07	3.964e+05	0.770	49.9
172.0	201.0	29.0	4.853e+02	1.904e-05	5.605e+06	1.933e+05	0.865	49.4
201.0	236.6	35.6	2.125e+02	1.330e-05	2.425e+06	6.810e+04	0.931	48.8
236.6	276.2	39.6	7.360e+01	7.321e-06	8.295e+05	2.095e+04	0.968	48.2
276.2	323.7	47.5	2.233e+01	3.505e-06	2.481e+05	5.222e+03	0.985	47.5
323.7	377.8	54.1	7.273e+00	1.798e-06	7.960e+04	1.471e+03	0.994	46.8
377.8	439.9	62.1	2.039e+00	7.866e-07	2.198e+04	3.540e+02	0.998	46.1
439.9	512.4	72.5	4.317e-01	2.585e-07	4.573e+03	6.308e+01	0.999	45.3
512.4	596.9	84.5	8.272e-02	7.690e-08	8.608e+02	1.019e+01	1.000	44.5
596.9	694.6	97.7	2.563e-02	3.692e-08	2.619e+02	2.680e+00	1.000	43.7
694.6	808.0	113.4	0.000e+00	0.000e+00	0.000e+00	0.000e+00	1.000	42.9
808.0	938.7	130.7	0.000e+00	0.000e+00	0.000e+00	0.000e+00	1.000	42.0
938.7	1091.8	153.1	0.000e+00	0.000e+00	0.000e+00	0.000e+00	1.000	41.1
1091.8	1267.3	175.5	0.000e+00	0.000e+00	0.000e+00	0.000e+00	1.000	40.3
1267.3	1471.8	204.5	0.000e+00	0.000e+00	0.000e+00	0.000e+00	1.000	39.3
			1.546e+06	2.005e-04	1.984e+10			

$D_{v0.1} = 29$ $D_{v0.5} = 111$ $D_{v0.9} = 220$ Mass balance closure 63%

Atomizer: SERG/ONAIR Rotational Rate (RPM): 14000 Surface tension 30 dynes/cm
Wind Speed (m/s): 49.2 PMS FSSP-100 & OAP-260X
Flow Rate 2.1 l/min Fluid: 50% isopropanol in glycol.
PMS Filename: T860220 de-icing fluid

Upper Diameter (microns)	Mid-range Diameter (Microns)	No. of Drops per sec. per micron (N_i / DD)	Cum. Vol. Frac. CVF	Cum. No. Frac. CNF
5.73	4.06	1138e+07	0.001	0.135
19.1	7.44	1318E+07	0.005	0.290
12.6	10.9	1104E+07	0.018	0.421
16.0	14.3	928E+07	0.042	0.531
19.5	17.8	782E+07	0.080	0.623
22.9	21.2	704E+07	0.138	0.706
26.3	24.6	638E+07	0.221	0.782
29.8	28.1	570E+07	0.331	0.849
33.2	31.5	466E+07	0.459	0.904
36.6	34.9	364E+07	0.579	0.943
40.1	38.4	210E+07	0.683	0.967
43.5	41.8	1348E+06	0.769	0.983
47.0	45.2	722E+06	0.827	0.992
50.4	48.7	394E+06	0.867	0.997
53.8	52.1	214E+06	0.893	0 .999
61.6	55.9	310E+05	0.904	0 .999
73.0	67.3	1768E+04	0.919	1.000
84.3	78.7	1186E+04	0.936	1.000
95.8	90.1	554E+04	0.948	1.000
108.2	102.0	372E+04	0.960	1.000
118.8	113.5	290E+04	0.972	1.000
130.0	124.4	1622E+03	0.981	1.000
141.5	135.8	848E+03	0.987	1.000
152.9	147.2	518E+03	0.992	1.000
164.4	158.6	232E+03	0.994	1.000
175.8	170.1	1318E+02	0.996	1.000
187.4	181.6	838E+02	0.998	1.000
199.0	193.2	476E+02	0.999	1.000
211	204.8	274E+02	0.999	1.000
222	216.4	148E+01	1.000	1.000
238	228	318E+01	1.000	1.000
245	240	816	1.000	1.000
257	251	836	1.000	1.000
269	263	0	1.000	1.000
280	274	0	1.000	1.000

$D_{v0.1} = 21$ $D_{v0.5} = 31$ $D_{v0.9} = 59$ (microns)

Atomizer: AU4000
Wind Speed (m/s): 49.2
Flow Rate 7.8 l/min.
PMS Filename: T860715C

Rotational Rate (RPM): 9000
PMS FSSP-100 & OAP-260X
Fluid: 20% isopropanol in water

Surface tension 40 dynes/cm

(m^3/s)

Upper Diameter (microns)	Mid-range Diameter (microns)	No. Of Drops per sec. Per micron (N_i /DD)	Cum. Vol. Frac. CVF	Cum. No. Frac. CNF
5.35	3.74	664E+07	0.000	0.121
8.55	6.95	462E+07	0.000	0.205
11.8	10.2	504E+07	0.002	0.297
15.0	13.4	540E+07	0.004	0.395
18.2	16.6	430E+07	0.009	0.473
21.4	19.8	380E+07	0.015	0.542
24.6	23.0	352E+07	0.025	0.606
27.8	26.2	360E+07	0.039	0.672
31.0	29.4	350E+07	0.059	0.736
34.2	32.6	322E+07	0.084	0.794
37.4	35.8	278E+07	0.112	0.845
40.6	39.0	238E+07	0.143	0.888
43.8	42.2	202E+07	0.177	0.925
47.0	45.4	1678E+06	0.212	0.956
50.2	48.6	1362E+06	0.247	0.981
61.6	55.9	1982E+05	0.274	0.984
73.0	67.3	224E+06	0.327	0.988
84.3	78.7	234E+06	0.417	0.993
95.8	90.1	1312E+05	0.493	0.995
108.2	102.0	812E+05	0.568	0.996
118.8	113.5	658E+05	0.638	0.998
130.0	124.4	438E+05	0.703	0.999
141.5	135.8	378E+05	0.759	0.999
152.9	147.2	1900E+04	0.806	1.000
164.4	158.6	1214E+04	0.845	1.000
175.8	170.1	808E+04	0.876	1.000
187.4	181.6	508E+04	0.901	1.000
199.0	193.2	404E+04	0.924	1.000
210	205	282E+04	0.943	1.000
222	216	206E+04	0.960	1.000
234	228	966E+03	0.969	1.000
245	240	836E+03	0.978	1.000
257	251	398E+03	0.983	1.000
269	263	354E+03	0.989	1.000
280	274	728E+02	0.990	1.000
292	286	1774E+02	0.993	1.000
303	298	1150E+02	0.996	1.000
315	309	1084E+02	0.998	1.000
327	321	304E+02	0.999	1.000
338	332	208E+02	1.000	1.000
350	344	0	1.000	1.000
361	356	5540	1.000	1.000
373	367	0	1.000	1.000

$D_{v0.1}$ = 36 $D_{v0.5}$ = 96 $D_{v0.9}$ = 187 (microns)

CP Nozzle
Deflector Plate Angle (degrees): 30
Orifice Size (inches): 0.062
Wind Speed (mph): 70
Pressure (psi): 25

Focal Length (mm): 800
Sweeps: 2500
Axial Location (m): 1.85
Fluid: 9% isopropanol in water
Malvern Filename: d16nov94

Surface tension 40 dynes/cm

Lower Diameter (microns)	Upper Diameter (microns)	DD	No. of Drops Noi (No.)	Volumetric Rate Fi (m^3/s)	Droplet Number Rate Ni (No./s)	Ni/DD	CVF	Calc. Droplet Velocity (m/s)
13.5	25.0	11.5	7.807e+02	2.727e-08	7.302e+06	6.350e+05	0.001	40.0
25.0	27.4	2.4	1.688e+02	1.487e-08	1.579e+06	6.577e+05	0.001	40.0
27.4	30.3	2.9	1.811e+02	2.130e-08	1.694e+06	5.841e+05	0.002	40.0
30.3	33.6	3.3	6.529e+01	1.043e-08	6.107e+05	1.851e+05	0.002	40.0
33.6	37.4	3.8	6.810e+01	1.485e-08	6.338e+05	1.668e+05	0.003	39.8
37.4	41.9	4.5	7.151e+01	2.161e-08	6.622e+05	1.472e+05	0.003	39.6
41.9	47.0	5.1	6.547e+01	2.774e-08	6.032e+05	1.183e+05	0.004	39.4
47.0	53.0	6.0	8.019e+01	4.786e-08	7.313e+05	1.219e+05	0.006	39.0
53.0	59.9	6.9	7.514e+01	6.421e-08	6.817e+05	9.880e+04	0.008	38.8
59.9	67.9	8.0	7.745e+01	9.501e-08	6.955e+05	8.694e+04	0.010	38.4
67.9	77.2	9.3	6.964e+01	1.241e-07	6.204e+05	6.671e+04	0.014	38.1
77.2	88.0	10.8	6.943e+01	1.806e-07	6.121e+05	5.667e+04	0.020	37.7
88.0	100.4	12.4	6.101e+01	2.329e-07	5.321e+05	4.291e+04	0.027	37.3
100.4	114.5	14.1	5.655e+01	3.178e-07	4.893e+05	3.470e+04	0.036	37.0
114.5	131.5	17.0	5.771e+01	4.812e-07	4.939e+05	2.905e+04	0.051	36.6
131.5	150.5	19.0	5.624e+01	6.988e-07	4.761e+05	2.506e+04	0.072	36.2
150.5	173.5	23.0	4.982e+01	9.258e-07	4.159e+05	1.808e+04	0.099	35.7
173.5	200.5	27.0	3.599e+01	1.017e-06	2.971e+05	1.100e+04	0.130	35.3
200.5	229.5	29.0	2.747e+01	1.167e-06	2.242e+05	7.730e+03	0.165	34.9
229.5	265.5	36.0	2.643e+01	1.693e-06	2.132e+05	5.923e+03	0.216	34.5
265.5	305.5	40.0	2.109e+01	2.043e-06	1.677e+05	4.192e+03	0.277	34.0
305.5	353.5	48.0	1.857e+01	2.724e-06	1.454e+05	3.030e+03	0.359	33.5
353.5	408.5	55.0	1.387e+01	3.109e-06	1.074e+05	1.952e+03	0.453	33.1
408.5	470.5	62.0	9.690e+00	3.283e-06	7.387e+04	1.191e+03	0.552	32.6
470.5	544.5	74.0	8.324e+00	4.290e-06	6.268e+04	8.470e+02	0.681	32.2
544.5	629.5	85.0	5.191e+00	4.075e-06	3.848e+04	4.527e+02	0.803	31.7
629.5	728.5	99.0	2.053e+00	2.447e-06	1.493e+04	1.508e+02	0.877	31.1
728.5	842.5	114.0	7.092e-01	1.284e-06	5.058e+03	4.437e+01	0.915	30.5
842.5	974.5	132.0	3.458e-01	9.523e-07	2.425e+03	1.837e+01	0.944	30.0
974.5	1129.5	155.0	1.774e-01	7.460e-07	1.224e+03	7.895e+00	0.966	29.5
1129.5	1306.5	177.0	9.303e-02	5.969e-07	6.309e+02	3.564e+00	0.984	29.0
1306.5	1513.5	207.0	5.353e-02	5.217e-07	3.555e+02	1.717e+00	1.000	28.4
Total			2.224e+03	3.325e-05	2.019e+07			

$D_{v0.1}$ = 174 $D_{v0.5}$ = 438 $D_{v0.9}$ = 798 (microns)

Mass balance closure 100%

CP Nozzle
Deflector Plate Angle (degrees): 90
Orifice Size (inches): 0.062
Wind Speed (mph): 70
Pressure (psi): 25
Malvern Filename: d18nov94

Focal Length (mm): 800
Sweeps: 1000
Axial Location (m): 0.9
Fluid: 9% isopropanol in water

Surface tension 40 dynes/cm

Lower Diameter (microns)	Upper Diameter (microns)	DD	No. of Drops Noi (No.)	Volumetric Rate Fi (m^3/s)	Droplet Number Rate Ni (No./s)	Ni/DD	CVF	Calc. Droplet Velocity (m/s)
13.5	25.0	11.5	2.654e+03	9.270e-08	2.482e+07	2.158e+06	0.003	40.0
25.0	27.4	2.4	5.995e+02	5.280e-08	5.607e+06	2.336e+06	0.005	40.0
27.4	30.3	2.9	5.377e+02	6.324e-08	5.030e+06	1.734e+06	0.007	40.0
30.3	33.6	3.3	4.729e+02	7.553e-08	4.423e+06	1.340e+06	0.010	40.0
33.6	37.4	3.8	4.290e+02	9.353e-08	3.993e+06	1.051e+06	0.013	39.8
37.4	41.9	4.5	4.198e+02	1.269e-07	3.887e+06	8.638e+05	0.017	39.6
41.9	47.0	5.1	3.504e+02	1.485e-07	3.229e+06	6.330e+05	0.022	39.4
47.0	53.0	6.0	2.844e+02	1.698e-07	2.594e+06	4.323e+05	0.028	39.0
53.0	59.9	6.9	2.577e+02	2.202e-07	2.338e+06	3.389e+05	0.035	38.8
59.9	67.9	8.0	2.433e+02	2.985e-07	2.185e+06	2.731e+05	0.045	38.4
67.9	77.2	9.3	2.202e+02	3.922e-07	1.962e+06	2.109e+05	0.058	38.1
77.2	88.0	10.8	2.501e+02	6.505e-07	2.205e+06	2.041e+05	0.080	37.7
88.0	100.4	12.4	2.088e+02	7.971e-07	1.821e+06	1.469e+05	0.106	37.3
100.4	114.5	14.1	1.663e+02	9.319e-07	1.435e+06	1.018e+05	0.138	36.9
114.5	131.5	17.0	1.266e+02	1.053e-06	1.081e+06	6.356e+04	0.173	36.5
131.5	150.5	19.0	1.016e+02	1.255e-06	8.549e+05	4.499e+04	0.215	36.0
150.5	173.5	23.0	8.963e+01	1.656e-06	7.441e+05	3.235e+04	0.270	35.5
173.5	200.5	27.0	7.366e+01	2.064e-06	6.029e+05	2.233e+04	0.339	35.0
200.5	229.5	29.0	5.312e+01	2.237e-06	4.298e+05	1.482e+04	0.414	34.6
229.5	265.5	36.0	5.520e+01	3.484e-06	4.389e+05	1.219e+04	0.531	34.0
265.5	305.5	40.0	4.429e+01	4.227e-06	3.469e+05	8.673e+03	0.672	33.5
305.5	353.5	48.0	3.011e+01	4.352e-06	2.324e+05	4.841e+03	0.818	33.0
353.5	408.5	55.0	1.606e+01	3.524e-06	1.217e+05	2.213e+03	0.936	32.4
408.5	470.5	62.0	4.130e+00	1.365e-06	3.071e+04	4.953e+02	0.981	31.8
470.5	544.5	74.0	8.417e-01	4.202e-07	6.140e+03	8.298e+01	0.995	31.2
544.5	629.5	85.0	8.791e-02	6.640e-08	6.269e+02	7.376e+00	0.998	30.5
629.5	728.5	99.0	2.675e-02	3.066e-08	1.870e+02	1.889e+00	0.999	29.9
728.5	842.5	114.0	4.493e-03	7.812e-09	3.079e+01	2.701e-01	0.999	29.3
842.5	974.5	132.0	2.594e-03	6.787e-09	1.729e+01	1.310e-01	0.999	28.5
974.5	1129.5	155.0	1.359e-03	5.386e-09	8.835e+00	5.700e-02	0.999	27.8
1129.5	1306.5	177.0	1.142e-03	6.823e-09	7.211e+00	4.074e-02	1.000	27.0
1306.5	1513.5	207.0	1.418e-03	1.280e-08	8.719e+00	4.212e-02	1.000	26.3
Total			7.689e+03	2.989e-05	7.042e+07			

$D_{v0.1}$ = 98 $D_{v0.5}$ = 256 $D_{v0.9}$ = 392 (microns) Mass balance closure 90%

CP Nozzle
Deflector Plate Angle (degrees): 45
Orifice Size (inches): 0.062
Wind Speed (mph): 70
Pressure (psi): 25

Focal Length (mm): 800
Sweeps: 1000
Axial Location (m): 0.9
Fluid: 9% isopropanol in water

Surface tension 40 dynes/cm

Malvern Filename: e18nov94

Lower Diameter (microns)	Upper Diameter (microns)	DD	No. of Drops Noi (No.)	Volumetric Rate Fi (m^3/s)	Droplet Number Rate Ni (Nc./s)	Ni/DD	CVF	Calc. Droplet Velocity (m/s)
13.5	25.0	11.5	1.026e+03	3.585e-08	9.599e+06	8.347e+05	0.001	40.0
25.0	27.4	2.4	3.771e+02	3.321e-08	3.527e+06	1.469e+06	0.002	40.0
27.4	30.3	2.9	2.941e+02	3.459e-08	2.751e+06	9.487e+05	0.003	40.0
30.3	33.6	3.3	3.066e+02	4.897e-08	2.867e+06	8.689e+05	0.004	40.0
33.6	37.4	3.8	3.253e+02	7.092e-08	3.027e+06	7.967e+05	0.006	39.8
37.4	41.9	4.5	2.830e+02	8.554e-08	2.621e+06	5.824e+05	0.009	39.6
41.9	47.0	5.1	2.759e+02	1.169e-07	2.542e+06	4.984e+05	0.012	39.4
47.0	53.0	6.0	2.320e+02	1.385e-07	2.116e+06	3.527e+05	0.016	39.0
53.0	59.9	6.9	1.940e+02	1.658e-07	1.760e+06	2.551e+05	0.021	38.8
59.9	67.9	8.0	1.876e+02	2.301e-07	1.685e+06	2.106e+05	0.027	38.4
67.9	77.2	9.3	1.985e+02	3.537e-07	1.769e+06	1.902e+05	0.037	38.1
77.2	88.0	10.8	1.949e+02	5.069e-07	1.718e+06	1.591e+05	0.052	37.7
88.0	100.4	12.4	1.705e+02	6.507e-07	1.487e+06	1.199e+05	0.071	37.3
100.4	114.5	14.1	1.360e+02	7.624e-07	1.174e+06	8.324e+04	0.092	36.9
114.5	131.5	17.0	1.042e+02	8.666e-07	8.894e+05	5.232e+04	0.117	36.5
131.5	150.5	19.0	9.500e+01	1.174e-06	7.997e+05	4.209e+04	0.150	36.0
150.5	173.5	23.0	8.559e+01	1.582e-06	7.105e+05	3.089e+04	0.196	35.5
173.5	200.5	27.0	7.565e+01	2.120e-06	6.192e+05	2.293e+04	0.256	35.0
200.5	229.5	29.0	5.951e+01	2.506e-06	4.815e+05	1.660e+04	0.328	34.6
229.5	265.5	36.0	5.390e+01	3.402e-06	4.285e+05	1.190e+04	0.425	34.0
265.5	305.5	40.0	4.163e+01	3.973e-06	3.261e+05	8.152e+03	0.538	33.5
305.5	353.5	48.0	3.524e+01	5.094e-06	2.720e+05	5.666e+03	0.683	33.0
353.5	408.5	55.0	2.114e+01	4.637e-06	1.601e+05	2.911e+03	0.816	32.4
408.5	470.5	62.0	7.170e+00	2.370e-06	5.331e+04	8.599e+02	0.883	31.8
470.5	544.5	74.0	2.994e+00	1.495e-06	2.184e+04	2.951e+02	0.926	31.2
544.5	629.5	85.0	1.425e+00	1.077e-06	1.017e+04	1.196e+02	0.957	30.5
629.5	728.5	99.0	5.903e-01	6.765e-07	4.127e+03	4.169e+01	0.976	29.9
728.5	842.5	114.0	1.926e-01	3.349e-07	1.320e+03	1.157e+01	0.985	29.3
842.5	974.5	132.0	8.117e-02	2.124e-07	5.410e+02	4.098e+00	0.991	28.5
974.5	1129.5	155.0	3.566e-02	1.413e-07	2.318e+02	1.495e+00	0.995	27.8
1129.5	1306.5	177.0	1.481e-02	8.848e-08	9.352e+01	5.284e-01	0.998	27.0
1306.5	1513.5	207.0	7.687e-03	6.939e-08	4.727e+01	2.284e-01	1.000	26.3
Total			4.786e+03	3.505e-05	4.342e+07			

$D_{v0.1}$ = 120 $D_{v0.5}$ = 292 $D_{v0.9}$ = 500 (microns) Mass balance closure 105%

CP Nozzle
Deflector Plate Angle (degrees): 45
Orifice Size (inches): 0.125
Wind Speed (mph): 70
Pressure (psi): 25
Malvern Filename: d29nov94

Focal Length (mm): 800
Sweeps: 1000
Axial Location (m): 0.9
Fluid: 9% isopropanol in water

Surface tension 40 dynes/cm

Lower Diameter (microns)	Upper Diameter (microns)	DD	No. of Drops Noi (No.)	Volumetric Rate Fi (m^3/s)	Droplet Number Rate Ni (No./s)	Ni/DD	CVF	Calc. Droplet Velocity (m/s)
13.5	25.0	11.5	5.859e+03	2.047e-07	5.480e+07	4.765e+06	0.002	40.0
25.0	27.4	2.4	1.085e+03	9.554e-08	1.015e+07	4.228e+06	0.003	40.0
27.4	30.3	2.9	8.610e+02	1.013e-07	8.054e+06	2.777e+06	0.004	40.0
30.3	33.6	3.3	4.105e+02	6.556e-08	3.839e+06	1.163e+06	0.005	40.0
33.6	37.4	3.8	3.330e+02	7.260e-08	3.099e+06	8.156e+05	0.006	39.8
37.4	41.9	4.5	4.603e+02	1.391e-07	4.262e+06	9.472e+05	0.007	39.6
41.9	47.0	5.1	5.037e+02	2.134e-07	4.641e+06	9.100e+05	0.010	39.4
47.0	53.0	6.0	5.898e+02	3.521e-07	5.379e+06	8.965e+05	0.013	39.0
53.0	59.9	6.9	4.962e+02	4.240e-07	4.502e+06	6.524e+05	0.018	38.8
59.9	67.9	8.0	4.014e+02	4.924e-07	3.604e+06	4.505e+05	0.023	38.4
67.9	77.2	9.3	3.250e+02	5.789e-07	2.895e+06	3.113e+05	0.029	38.1
77.2	88.0	10.8	2.976e+02	7.741e-07	2.623e+06	2.429e+05	0.038	37.7
88.0	100.4	12.4	2.744e+02	1.048e-06	2.393e+06	1.930e+05	0.049	37.3
100.4	114.5	14.1	2.524e+02	1.415e-06	2.178e+06	1.545e+05	0.064	36.9
114.5	131.5	17.0	2.302e+02	1.914e-06	1.964e+06	1.155e+05	0.085	36.5
131.5	150.5	19.0	2.198e+02	2.715e-06	1.850e+06	9.737e+04	0.114	36.0
150.5	173.5	23.0	2.059e+02	3.804e-06	1.709e+06	7.431e+04	0.155	35.5
173.5	200.5	27.0	1.420e+02	3.978e-06	1.162e+06	4.303e+04	0.198	35.0
200.5	229.5	29.0	9.638e+01	4.058e-06	7.798e+05	2.689e+04	0.241	34.6
229.5	265.5	36.0	8.608e+01	5.433e-06	6.844e+05	1.901e+04	0.300	34.0
265.5	305.5	40.0	6.547e+01	6.249e-06	5.128e+05	1.282e+04	0.367	33.5
305.5	353.5	48.0	6.231e+01	9.006e-06	4.808e+05	1.002e+04	0.463	33.0
353.5	408.5	55.0	5.112e+01	1.121e-05	3.873e+05	7.041e+03	0.584	32.4
408.5	470.5	62.0	3.636e+01	1.202e-05	2.704e+05	4.361e+03	0.713	31.8
470.5	544.5	74.0	2.360e+01	1.178e-05	1.722e+05	2.327e+03	0.840	31.2
544.5	629.5	85.0	1.197e+01	9.040e-06	8.536e+04	1.004e+03	0.937	30.5
629.5	728.5	99.0	3.594e+00	4.119e-06	2.513e+04	2.538e+02	0.981	29.9
728.5	842.5	114.0	6.486e-01	1.128e-06	4.443e+03	3.898e+01	0.993	29.3
842.5	974.5	132.0	1.451e-01	3.796e-07	9.669e+02	7.325e+00	0.997	28.5
974.5	1129.5	155.0	4.342e-02	1.721e-07	2.822e+02	1.821e+00	0.999	27.8
1129.5	1306.5	177.0	1.130e-02	6.749e-08	7.134e+01	4.030e-01	1.000	27.0
1306.5	1513.5	207.0	1.935e-03	1.746e-08	1.190e+01	5.748e-02	1.000	26.3
Total			1.338e+04	9.307e-05	1.225e+08			

$D_{v0.1}$ = 141 $D_{v0.5}$ = 370 $D_{v0.9}$ = 597 (microns) Mass balance closure 93%

CP Nozzle
Deflector Plate Angle (degrees): 90
Orifice Size (inches): 0.125
Wind Speed (mph): 70
Pressure (psi): 25
Malvern Filename: e29nov94

Focal Length (mm): 800
Sweeps: 1000
Axial Location (m): 0.9
Fluid: 9% isopropanol in water

Surface tension 40 dynes/cm

Lower Diameter (microns)	Upper Diameter (microns)	DD	No. of Drops Noi (No.)	Volumetric Rate Fi (m^3/s)	Droplet Number Rate Ni (No./s)	Ni/DD	CVF	Calc. Droplet Velocity (m/s)
13.5	25.0	11.5	9.396e+03	3.283e-07	8.789e+07	7.642e+06	0.003	40.0
25.0	27.4	2.4	1.995e+03	1.758e-07	1.866e+07	7.777e+06	0.005	40.0
27.4	30.3	2.9	1.480e+03	1.740e-07	1.384e+07	4.773e+06	0.007	40.0
30.3	33.6	3.3	1.062e+03	1.696e-07	9.931e+06	3.009e+06	0.009	40.0
33.6	37.4	3.8	8.630e+02	1.881e-07	8.031e+06	2.114e+06	0.011	39.8
37.4	41.9	4.5	1.166e+03	3.525e-07	1.080e+07	2.400e+06	0.015	39.6
41.9	47.0	5.1	1.147e+03	4.858e-07	1.056e+07	2.071e+06	0.020	39.4
47.0	53.0	6.0	9.231e+02	5.510e-07	8.418e+06	1.403e+06	0.025	39.0
53.0	59.9	6.9	7.303e+02	6.240e-07	6.626e+06	9.602e+05	0.032	38.8
59.9	67.9	8.0	6.393e+02	7.843e-07	5.741e+06	7.176e+05	0.040	38.4
67.9	77.2	9.3	6.430e+02	1.145e-06	5.728e+06	6.160e+05	0.052	38.1
77.2	88.0	10.8	6.108e+02	1.589e-06	5.384e+06	4.985e+05	0.069	37.7
88.0	100.4	12.4	5.132e+02	1.959e-06	4.476e+06	3.610e+05	0.089	37.3
100.4	114.5	14.1	4.283e+02	2.401e-06	3.696e+06	2.621e+05	0.115	36.9
114.5	131.5	17.0	3.369e+02	2.802e-06	2.876e+06	1.692e+05	0.144	36.5
131.5	150.5	19.0	2.696e+02	3.331e-06	2.269e+06	1.194e+05	0.179	36.0
150.5	173.5	23.0	2.262e+02	4.180e-06	1.878e+06	8.164e+04	0.223	35.5
173.5	200.5	27.0	1.663e+02	4.661e-06	1.361e+06	5.042e+04	0.272	35.0
200.5	229.5	29.0	1.239e+02	5.219e-06	1.003e+06	3.458e+04	0.326	34.6
229.5	265.5	36.0	1.115e+02	7.040e-06	8.868e+05	2.463e+04	0.400	34.0
265.5	305.5	40.0	8.787e+01	8.387e-06	6.883e+05	1.721e+04	0.488	33.5
305.5	353.5	48.0	8.587e+01	1.241e-05	6.627e+05	1.381e+04	0.619	33.0
353.5	408.5	55.0	6.176e+01	1.355e-05	4.679e+05	8.508e+03	0.761	32.4
408.5	470.5	62.0	3.111e+01	1.028e-05	2.313e+05	3.731e+03	0.869	31.8
470.5	544.5	74.0	1.300e+01	6.490e-06	9.482e+04	1.281e+03	0.937	31.2
544.5	629.5	85.0	4.748e+00	3.586e-06	3.386e+04	3.984e+02	0.974	30.5
629.5	728.5	99.0	1.517e+00	1.739e-06	1.061e+04	1.072e+02	0.992	29.9
728.5	842.5	114.0	2.718e-01	4.727e-07	1.863e+03	1.634e+01	0.997	29.3
842.5	974.5	132.0	2.575e-02	6.737e-08	1.716e+02	1.300e+00	0.998	28.5
974.5	1129.5	155.0	1.605e-02	6.361e-08	1.044e+02	6.732e-01	0.999	27.8
1129.5	1306.5	177.0	9.613e-03	5.742e-08	6.069e+01	3.429e-01	0.999	27.0
1306.5	1513.5	207.0	6.196e-03	5.593e-08	3.810e+01	1.841e-01	1.000	26.3
Total			2.312e+04	9.532e-05	2.123e+08			

$D_{v0.1}$ = 106 $D_{v0.5}$ = 310 $D_{v0.9}$ = 504 (microns) Mass balance closure 95%

CP Nozzle
Deflector Plate Angle (degrees): 45
Orifice Size (inches): 0.078
Wind Speed (mph): 70
Pressure (psi): 25

Focal Length (mm): 800
Sweeps: 1000
Axial Location (m): 0.9
Fluid: 9% isopropanol in water
Malvern Filename: f18nov94

Surafce tension 40 dynes/cm

Lower Diameter (microns)	Upper Diameter (microns)	DD	No. of Drops Noi (No.)	Volumetric Rate Fi (m^3/s)	Droplet Number Rate Ni (No./s)	Ni/DD	CVF	Calc. Droplet Velocity (m / s)
13.5	25.0	11.5	1.979e+03	6.912e-08	1.851e+07	1.609e+06	0.001	40.0
25.0	27.4	2.4	4.349e+02	3.830e-08	4.068e+06	1.695e+06	0.002	40.0
27.4	30.3	2.9	4.525e+02	5.322e-08	4.233e+06	1.459e+06	0.003	40.0
30.3	33.6	3.3	3.588e+02	5.731e-08	3.356e+06	1.017e+06	0.004	40.0
33.6	37.4	3.8	3.277e+02	7.144e-08	3.050e+06	8.026e+05	0.006	39.8
37.4	41.9	4.5	3.749e+02	1.133e-07	3.471e+06	7.714e+05	0.008	39.6
41.9	47.0	5.1	3.417e+02	1.448e-07	3.148e+06	6.173e+05	0.011	39.4
47.0	53.0	6.0	2.913e+02	1.739e-07	2.657e+06	4.428e+05	0.014	39.0
53.0	59.9	6.9	2.454e+02	2.097e-07	2.227e+06	3.227e+05	0.018	38.8
59.9	67.9	8.0	2.391e+02	2.933e-07	2.147e+06	2.683e+05	0.024	38.4
67.9	77.2	9.3	2.748e+02	4.896e-07	2.449e+06	2.633e+05	0.034	38.1
77.2	88.0	10.8	2.871e+02	7.470e-07	2.531e+06	2.344e+05	0.048	37.7
88.0	100.4	12.4	2.658e+02	1.015e-06	2.318e+06	1.869e+05	0.068	37.3
100.4	114.5	14.1	2.197e+02	1.232e-06	1.896e+06	1.345e+05	0.092	36.9
114.5	131.5	17.0	1.683e+02	1.399e-06	1.436e+06	8.448e+04	0.120	36.5
131.5	150.5	19.0	1.358e+02	1.678e-06	1.143e+06	6.016e+04	0.153	36.0
150.5	173.5	23.0	1.165e+02	2.153e-06	9.671e+05	4.205e+04	0.195	35.5
173.5	200.5	27.0	9.752e+01	2.733e-06	7.981e+05	2.956e+04	0.249	35.0
200.5	229.5	29.0	7.503e+01	3.159e-06	6.071e+05	2.093e+04	0.311	34.6
229.5	265.5	36.0	6.732e+01	4.249e-06	5.352e+05	1.487e+04	0.394	34.0
265.5	305.5	40.0	4.887e+01	4.664e-06	3.828e+05	9.570e+03	0.486	33.5
305.5	353.5	48.0	4.580e+01	6.620e-06	3.534e+05	7.363e+03	0.616	33.0
353.5	408.5	55.0	3.279e+01	7.193e-06	2.484e+05	4.516e+03	0.757	32.4
408.5	470.5	62.0	1.339e+01	4.424e-06	9.953e+04	1.605e+03	0.844	31.8
470.5	544.5	74.0	6.154e+00	3.073e-06	4.490e+04	6.067e+02	0.904	31.2
544.5	629.5	85.0	3.035e+00	2.292e-06	2.164e+04	2.546e+02	0.949	30.5
629.5	728.5	99.0	1.351e+00	1.549e-06	9.449e+03	9.545e+01	0.980	29.9
728.5	842.5	114.0	2.723e-01	4.734e-07	1.865e+03	1.636e+01	0.989	29.3
842.5	974.5	132.0	8.203e-02	2.146e-07	5.466e+02	4.141e+00	0.993	28.5
974.5	1129.5	155.0	3.680e-02	1.458e-07	2.392e+02	1.543e+00	0.996	27.8
1129.5	1306.5	177.0	1.789e-02	1.069e-07	1.130e+02	6.382e-01	0.998	27.0
1306.5	1513.5	207.0	1.003e-02	9.054e-08	6.168e+01	2.980e-01	1.000	26.3
Total			6.905e+03	5.092e-05	6.271e+07			

$D_{v0.1}$ = 119 $D_{v0.5}$ = 328 $D_{v0.9}$ = 540 (microns) Mass balance closure 99%

CP Nozzle
Deflector Plate Angle (degrees): 90
Orifice Size (inches): 0.078
Wind Speed (mph): 70
Pressure (psi): 25
Malvern Filename: d22nov94
Focal Length (mm): 800 Surface tension 40 dynes/cm
Sweeps: 1000
Axial Location (m): 0.9
Fluid: 9% isopropanol in water

Lower Diameter (microns)	Upper Diameter (microns)	DD	No. of Drops Noi (No.)	Volumetric Rate Fi (m^3/s)	Droplet Number Rate Ni (No./s)	Ni/DD	CVF	Calc. Droplet Velocity (m / s)
13.5	25.0	11.5	2.630e+03	9.187e-08	2.460e+07	2.139e+06	0.002	40.0
25.0	27.4	2.4	6.669e+02	5.874e-08	6.238e+06	2.599e+06	0.003	40.0
27.4	30.3	2.9	4.811e+02	5.658e-08	4.500e+06	1.552e+06	0.005	40.0
30.3	33.6	3.3	3.661e+02	5.848e-08	3.425e+06	1.038e+06	0.006	40.0
33.6	37.4	3.8	4.443e+02	9.687e-08	4.135e+06	1.088e+06	0.008	39.8
37.4	41.9	4.5	4.520e+02	1.366e-07	4.186e+06	9.301e+05	0.011	39.6
41.9	47.0	5.1	3.649e+02	1.546e-07	3.362e+06	6.592e+05	0.015	39.4
47.0	53.0	6.0	3.330e+02	1.987e-07	3.036e+06	5.061e+05	0.020	39.0
53.0	59.9	6.9	2.827e+02	2.415e-07	2.565e+06	3.717e+05	0.025	38.8
59.9	67.9	8.0	2.856e+02	3.504e-07	2.565e+06	3.206e+05	0.033	38.4
67.9	77.2	9.3	2.886e+02	5.141e-07	2.571e+06	2.765e+05	0.045	38.1
77.2	88.0	10.8	3.740e+02	9.729e-07	3.297e+06	3.053e+05	0.067	37.7
88.0	100.4	12.4	2.715e+02	1.037e-06	2.368e+06	1.910e+05	0.091	37.3
100.4	114.5	14.1	1.960e+02	1.099e-06	1.691e+06	1.200e+05	0.116	36.9
114.5	131.5	17.0	1.600e+02	1.331e-06	1.366e+06	8.033e+04	0.147	36.5
131.5	150.5	19.0	1.313e+02	1.622e-06	1.105e+06	5.816e+04	0.184	36.0
150.5	173.5	23.0	1.165e+02	2.153e-06	9.671e+05	4.205e+04	0.233	35.5
173.5	200.5	27.0	1.063e+02	2.979e-06	8.702e+05	3.223e+04	0.301	35.0
200.5	229.5	29.0	8.133e+01	3.424e-06	6.580e+05	2.269e+04	0.380	34.6
229.5	265.5	36.0	7.623e+01	4.811e-06	6.060e+05	1.683e+04	0.490	34.0
265.5	305.5	40.0	5.938e+01	5.668e-06	4.652e+05	1.163e+04	0.620	33.5
305.5	353.5	48.0	4.703e+01	6.798e-06	3.629e+05	7.561e+03	0.776	33.0
353.5	408.5	55.0	2.854e+01	6.262e-06	2.163e+05	3.932e+03	0.919	32.4
408.5	470.5	62.0	8.030e+00	2.654e-06	5.971e+04	9.631e+02	0.980	31.8
470.5	544.5	74.0	1.000e+00	4.993e-07	7.296e+03	9.859e+01	0.992	31.2
544.5	629.5	85.0	3.667e-01	2.769e-07	2.615e+03	3.077e+01	0.998	30.5
629.5	728.5	99.0	4.427e-02	5.074e-08	3.095e+02	3.127e+00	0.999	29.9
728.5	842.5	114.0	6.547e-03	1.138e-08	4.486e+01	3.935e-01	0.999	29.3
842.5	974.5	132.0	4.415e-03	1.155e-08	2.942e+01	2.229e-01	1.000	28.5
974.5	1129.5	155.0	1.314e-03	5.206e-09	8.539e+00	5.509e-02	1.000	27.8
1129.5	1306.5	177.0	8.464e-04	5.056e-09	5.344e+00	3.019e-02	1.000	27.0
1306.5	1513.5	207.0	8.184e-04	7.387e-09	5.033e+00	2.431e-02	1.000	26.3
Total			8.253e+03	4.364e-05	7.522e+07			

$D_{v0.1}$ = 105 $D_{v0.5}$ = 269 $D_{v0.9}$ = 401 (microns) Mass balance closure 84%

CP Nozzle
Deflector Plate Angle (degrees):30 Focal Length (mm): 800 Surface tension 40 dynes/cm
Orifice Size (inches): 0.078 Sweeps: 1000
Wind Speed (mph): 70 Axial Location (m): 0.9
Pressure (psi): 25 Fluid: 9% isopropanol in water
Malvern Filename: e22nov94

Lower Diameter (microns)	Upper Diameter (microns)	DD	No. of Drops Noi (No.)	Volumetric Rate Fi (m^3/s)	Droplet Number Rate Ni (No./s)	Ni/DD	CVF	Calc. Droplet Velocity (m/s)
13.5	25.0	11.5	1.673e+03	5.843e-08	1.564e+07	1.360e+06	0.001	40.0
25.0	27.4	2.4	2.064e+02	1.818e-08	1.930e+06	8.042e+05	0.002	40.0
27.4	30.3	2.9	2.343e+02	2.756e-08	2.192e+06	7.558e+05	0.002	40.0
30.3	33.6	3.3	3.085e+01	4.927e-09	2.885e+05	8.743e+04	0.002	40.0
33.6	37.4	3.8	2.249e+01	4.903e-09	2.093e+05	5.508e+04	0.002	39.8
37.4	41.9	4.5	6.845e+01	2.069e-08	6.338e+05	1.408e+05	0.003	39.6
41.9	47.0	5.1	9.992e+01	4.233e-08	9.206e+05	1.805e+05	0.004	39.4
47.0	53.0	6.0	1.296e+02	7.735e-08	1.182e+06	1.970e+05	0.005	39.0
53.0	59.9	6.9	1.007e+02	8.601e-08	9.132e+05	1.323e+05	0.007	38.8
59.9	67.9	8.0	8.698e+01	1.067e-07	7.810e+05	9.762e+04	0.009	38.4
67.9	77.2	9.3	7.705e+01	1.373e-07	6.865e+05	7.381e+04	0.012	38.1
77.2	88.0	10.8	8.439e+01	2.195e-07	7.439e+05	6.888e+04	0.017	37.7
88.0	100.4	12.4	7.702e+01	2.940e-07	6.718e+05	5.418e+04	0.023	37.3
100.4	114.5	14.1	7.334e+01	4.110e-07	6.328e+05	4.488e+04	0.031	36.9
114.5	131.5	17.0	7.259e+01	6.037e-07	6.195e+05	3.644e+04	0.044	36.5
131.5	150.5	19.0	6.456e+01	7.977e-07	5.435e+05	2.860e+04	0.060	36.0
150.5	173.5	23.0	5.970e+01	1.103e-06	4.956e+05	2.155e+04	0.083	35.5
173.5	200.5	27.0	4.232e+01	1.186e-06	3.464e+05	1.283e+04	0.107	35.0
200.5	229.5	29.0	3.529e+01	1.486e-06	2.855e+05	9.846e+03	0.138	34.6
229.5	265.5	36.0	3.835e+01	2.420e-06	3.049e+05	8.469e+03	0.188	34.0
265.5	305.5	40.0	3.094e+01	2.954e-06	2.424e+05	6.060e+03	0.248	33.5
305.5	353.5	48.0	2.283e+01	3.299e-06	1.761e+05	3.670e+03	0.316	33.0
353.5	408.5	55.0	1.760e+01	3.860e-06	1.333e+05	2.424e+03	0.396	32.4
408.5	470.5	62.0	1.631e+01	5.392e-06	1.213e+05	1.957e+03	0.507	31.8
470.5	544.5	74.0	1.585e+01	7.912e-06	1.156e+05	1.562e+03	0.670	31.2
544.5	629.5	85.0	9.809e+00	7.409e-06	6.996e+04	8.231e+02	0.823	30.5
629.5	728.5	99.0	3.330e+00	3.816e-06	2.328e+04	2.352e+02	0.901	29.9
728.5	842.5	114.0	8.694e-01	1.512e-06	5.957e+03	5.225e+01	0.933	29.3
842.5	974.5	132.0	4.305e-01	1.127e-06	2.869e+03	2.174e+01	0.956	28.5
974.5	1129.5	155.0	2.196e-01	8.701e-07	1.427e+03	9.208e+00	0.974	27.8
1129.5	1306.5	177.0	1.148e-01	6.859e-07	7.250e+02	4.096e+00	0.988	27.0
1306.5	1513.5	207.0	6.535e-02	5.899e-07	4.019e+02	1.941e+00	1.000	26.3
Total			3.395e+03	4.853e-05	3.092e+07			

$D_{v0.1} = 192$ $D_{v0.5} = 467$ $D_{v0.9} = 727$ (microns) Mass balance closure 94%

CP Nozzle
Deflector Plate Angle (degrees): 30
Orifice Size (inches): 0.125
Wind Speed (mph): 70
Pressure (psi): 25
Malvern Filename: f22nov94
Focal Length (mm): 800
Sweeps: 1000
Axial Location (m): 0.9
Fluid: 9% isopropanol in water
Surface tension 40 dynes/cm

Lower Diameter (microns)	Upper Diameter (microns)	DD	No. of Drops Noi (No.)	Volumetric Rate Fi (m^3/s)	Droplet Number Rate Ni (No./s)	Ni/DD	CVF	Calc. Droplet Velocity (m/s)
13.5	25.0	11.5	1.329e+03	4.644e-08	1.243e+07	1.081e+06	0.000	40.0
25.0	27.4	2.4	4.692e+02	4.133e-08	4.389e+06	1.829e+06	0.001	40.0
27.4	30.3	2.9	2.334e+02	2.744e-08	2.183e+06	7.526e+05	0.001	40.0
30.3	33.6	3.3	2.191e+02	3.500e-08	2.049e+06	6.210e+05	0.002	40.0
33.6	37.4	3.8	1.915e+02	4.176e-08	1.783e+06	4.691e+05	0.002	39.8
37.4	41.9	4.5	3.791e+02	1.146e-07	3.511e+06	7.802e+05	0.003	39.6
41.9	47.0	5.1	2.888e+02	1.224e-07	2.661e+06	5.217e+05	0.005	39.4
47.0	53.0	6.0	2.019e+02	1.205e-07	1.841e+06	3.069e+05	0.006	39.0
53.0	59.9	6.9	1.399e+02	1.195e-07	1.269e+06	1.839e+05	0.007	38.8
59.9	67.9	8.0	1.665e+02	2.043e-07	1.495e+06	1.869e+05	0.009	38.4
67.9	77.2	9.3	1.843e+02	3.284e-07	1.642e+06	1.766e+05	0.013	38.1
77.2	88.0	10.8	1.740e+02	4.526e-07	1.534e+06	1.420e+05	0.017	37.7
88.0	100.4	12.4	1.426e+02	5.443e-07	1.244e+06	1.003e+05	0.023	37.3
100.4	114.5	14.1	1.052e+02	5.897e-07	9.079e+05	6.439e+04	0.029	36.9
114.5	131.5	17.0	1.169e+02	9.726e-07	9.982e+05	5.872e+04	0.040	36.5
131.5	150.5	19.0	1.196e+02	1.477e-06	1.006e+06	5.297e+04	0.055	36.0
150.5	173.5	23.0	1.072e+02	1.980e-06	8.897e+05	3.868e+04	0.076	35.5
173.5	200.5	27.0	7.787e+01	2.182e-06	6.373e+05	2.360e+04	0.099	35.0
200.5	229.5	29.0	6.498e+01	2.736e-06	5.258e+05	1.813e+04	0.128	34.6
229.5	265.5	36.0	8.475e+01	5.349e-06	6.738e+05	1.872e+04	0.185	34.0
265.5	305.5	40.0	6.222e+01	5.939e-06	4.874e+05	1.219e+04	0.247	33.5
305.5	353.5	48.0	3.496e+01	5.053e-06	2.697e+05	5.620e+03	0.301	33.0
353.5	408.5	55.0	2.207e+01	4.841e-06	1.672e+05	3.040e+03	0.352	32.4
408.5	470.5	62.0	2.225e+01	7.354e-06	1.654e+05	2.669e+03	0.429	31.8
470.5	544.5	74.0	2.521e+01	1.259e-05	1.839e+05	2.485e+03	0.562	31.2
544.5	629.5	85.0	1.843e+01	1.392e-05	1.315e+05	1.547e+03	0.709	30.5
629.5	728.5	99.0	9.557e+00	1.095e-05	6.682e+04	6.750e+02	0.825	29.9
728.5	842.5	114.0	4.057e+00	7.054e-06	2.780e+04	2.438e+02	0.899	29.3
842.5	974.5	132.0	1.720e+00	4.501e-06	1.146e+04	8.684e+01	0.947	28.5
974.5	1129.5	155.0	6.706e-01	2.657e-06	4.359e+03	2.812e+01	0.975	27.8
1129.5	1306.5	177.0	2.478e-01	1.480e-06	1.565e+03	8.839e+00	0.991	27.0
1306.5	1513.5	207.0	9.855e-02	8.896e-07	6.061e+02	2.928e+00	1.000	26.3
	Total		4.998e+03	9.472e-05	4.519e+07			

$D_{v0.1} = 203$ $D_{v0.5} = 511$ $D_{v0.9} = 846$ (microns) Mass balance closure 91%

Atomizer: D10-45
Wind Speed (m/s): 22
Flowrate (l/min): 3.4
Malvern Filename: e03dec94

Orientation (degrees): 90
Focal Length (mm): 800
Sweeps: 1000
Axial Location (m): 0.9

Fluid: 29% isopropanol in water
Surface tension 31 dynes/cm

Lower Diameter (microns)	Upper Diameter (microns)	DD	No. of Drops Noi (No.)	Volumetric Rate Fi (m^3/s)	Droplet Number Rate Ni (No./s)	Ni/DD	CVF	Calc. Droplet Velocity (m/s)
13.5	25.0	11.5	5.355e+03	1.143e-07	3.061e+07	2.662e+06	0.002	24.4
25.0	27.4	2.4	1.170e+03	6.299e-08	6.689e+06	2.787e+06	0.003	24.4
27.4	30.3	2.9	4.613e+02	3.316e-08	2.637e+06	9.093e+05	0.004	24.4
30.3	33.6	3.3	4.694e+02	4.582e-08	2.683e+06	8.130e+05	0.005	24.4
33.6	37.4	3.8	5.092e+02	6.784e-08	2.896e+06	7.621e+05	0.006	24.3
37.4	41.9	4.5	5.708e+02	1.054e-07	3.230e+06	7.178e+05	0.008	24.2
41.9	47.0	5.1	4.566e+02	1.182e-07	2.571e+06	5.041e+05	0.010	24.1
47.0	53.0	6.0	5.179e+02	1.889e-07	2.886e+06	4.810e+05	0.013	23.8
53.0	59.9	6.9	4.834e+02	2.524e-07	2.680e+06	3.884e+05	0.018	23.7
59.9	67.9	8.0	4.455e+02	3.340e-07	2.445e+06	3.056e+05	0.024	23.5
67.9	77.2	9.3	3.912e+02	4.259e-07	2.130e+06	2.290e+05	0.032	23.3
77.2	88.0	10.8	2.924e+02	4.648e-07	1.575e+06	1.458e+05	0.040	23.0
88.0	100.4	12.4	2.645e+02	6.170e-07	1.410e+06	1.137e+05	0.051	22.8
100.4	114.5	14.1	2.546e+02	8.721e-07	1.343e+06	9.522e+04	0.067	22.6
114.5	131.5	17.0	2.391e+02	1.215e-06	1.247e+06	7.337e+04	0.089	22.3
131.5	150.5	19.0	2.334e+02	1.763e-06	1.201e+06	6.320e+04	0.121	22.0
150.5	173.5	23.0	2.078e+02	2.347e-06	1.054e+06	4.583e+04	0.163	21.7
173.5	200.5	27.0	1.698e+02	2.907e-06	8.491e+05	3.145e+04	0.215	21.4
200.5	229.5	29.0	1.398e+02	3.598e-06	6.914e+05	2.384e+04	0.280	21.1
229.5	265.5	36.0	1.165e+02	4.493e-06	5.660e+05	1.572e+04	0.362	20.8
265.5	305.5	40.0	8.562e+01	4.994e-06	4.099e+05	1.025e+04	0.452	20.5
305.5	353.5	48.0	6.178e+01	5.457e-06	2.913e+05	6.069e+03	0.550	20.2
353.5	408.5	55.0	3.962e+01	5.312e-06	1.834e+05	3.335e+03	0.646	19.8
408.5	470.5	62.0	2.533e+01	5.117e-06	1.151e+05	1.857e+03	0.739	19.4
470.5	544.5	74.0	1.643e+01	5.015e-06	7.327e+04	9.902e+02	0.829	19.1
544.5	629.5	85.0	8.784e+00	4.054e-06	3.828e+04	4.504e+02	0.902	18.6
629.5	728.5	99.0	3.135e+00	2.196e-06	1.340e+04	1.353e+02	0.942	18.3
728.5	842.5	114.0	1.013e+00	1.076e-06	4.241e+03	3.720e+01	0.961	17.9
842.5	974.5	132.0	4.695e-01	7.507e-07	1.912e+03	1.449e+01	0.975	17.4
974.5	1129.5	155.0	2.281e-01	5.525e-07	9.063e+02	5.847e+00	0.985	17.0
1129.5	1306.5	177.0	1.247e-01	4.551e-07	4.810e+02	2.718e+00	0.993	16.5
1306.5	1513.5	207.0	6.904e-02	3.809e-07	2.595e+02	1.254e+00	1.000	16.1
Total			1.299e+04	5.539e-05	7.252e+07			

$D_{v0.1}$ = 138 $D_{v0.5}$ = 329 $D_{v0.9}$ = 627 (microns) Mass balance closure 98%

Atomizer: D10-45
Wind Speed (m/s): 36
Flowrate (l/min): 3.4
Malvern Filename: d03dec94

Orientation (degrees): 90
Focal Length (mm): 800
Sweeps: 1000
Axial Location (m): 0.9

Fluid: 29% isopropanol in water
Surface tension 31 dynes/cm

Lower Diameter (microns)	Upper Diameter (microns)	DD	No. of Drops Noi (No.)	Volumetric Rate Fi (m^3/s)	Droplet Number Rate Ni (No./s)	Ni/DD	CVF	Calc. Droplet Velocity (m/s)
13.5	25.0	11.5	4.564e+03	1.594e-07	4.269e+07	3.712e+06	0.003	40.0
25.0	27.4	2.4	1.028e+03	9.050e-08	9.611e+06	4.004e+06	0.005	40.0
27.4	30.3	2.9	1.020e+03	1.200e-07	9.542e+06	3.290e+06	0.007	40.0
30.3	33.6	3.3	8.390e+02	1.340e-07	7.848e+06	2.378e+06	0.010	40.0
33.6	37.4	3.8	8.052e+02	1.755e-07	7.494e+06	1.972e+06	0.013	39.8
37.4	41.9	4.5	8.441e+02	2.551e-07	7.817e+06	1.737e+06	0.019	39.6
41.9	47.0	5.1	7.628e+02	3.232e-07	7.028e+06	1.378e+06	0.025	39.4
47.0	53.0	6.0	6.437e+02	3.842e-07	5.871e+06	9.784e+05	0.033	39.0
53.0	59.9	6.9	6.155e+02	5.260e-07	5.585e+06	8.094e+05	0.043	38.8
59.9	67.9	8.0	5.514e+02	6.764e-07	4.951e+06	6.189e+05	0.056	38.4
67.9	77.2	9.3	5.244e+02	9.341e-07	4.672e+06	5.023e+05	0.075	38.1
77.2	88.0	10.8	4.237e+02	1.102e-06	3.735e+06	3.458e+05	0.097	37.7
88.0	100.4	12.4	2.893e+02	1.105e-06	2.524e+06	2.035e+05	0.119	37.3
100.4	114.5	14.1	2.105e+02	1.180e-06	1.816e+06	1.288e+05	0.142	36.9
114.5	131.5	17.0	1.760e+02	1.464e-06	1.502e+06	8.836e+04	0.171	36.5
131.5	150.5	19.0	1.865e+02	2.305e-06	1.570e+06	8.264e+04	0.217	36.0
150.5	173.5	23.0	1.869e+02	3.454e-06	1.552e+06	6.746e+04	0.286	35.5
173.5	200.5	27.0	1.496e+02	4.191e-06	1.224e+06	4.533e+04	0.369	35.0
200.5	229.5	29.0	1.049e+02	4.415e-06	8.485e+05	2.926e+04	0.457	34.6
229.5	265.5	36.0	8.093e+01	5.108e-06	6.434e+05	1.787e+04	0.558	34.0
265.5	305.5	40.0	5.699e+01	5.440e-06	4.465e+05	1.116e+04	0.666	33.5
305.5	353.5	48.0	4.372e+01	6.319e-06	3.374e+05	7.028e+03	0.792	33.0
353.5	408.5	55.0	2.391e+01	5.245e-06	1.811e+05	3.293e+03	0.896	32.4
408.5	470.5	62.0	6.989e+00	2.310e-06	5.197e+04	8.382e+02	0.942	31.8
470.5	544.5	74.0	2.615e+00	1.306e-06	1.908e+04	2.578e+02	0.968	31.2
544.5	629.5	85.0	1.281e+00	9.676e-07	9.137e+03	1.075e+02	0.987	30.5
629.5	728.5	99.0	5.148e-01	5.900e-07	3.599e+03	3.636e+01	0.999	29.9
728.5	842.5	114.0	2.125e-02	3.695e-08	1.456e+02	1.277e+00	0.999	29.3
842.5	974.5	132.0	4.250e-03	1.112e-08	2.832e+01	2.146e-01	0.999	28.5
974.5	1129.5	155.0	2.737e-03	1.085e-08	1.779e+01	1.148e-01	1.000	27.8
1129.5	1306.5	177.0	1.764e-03	1.054e-08	1.114e+01	6.291e-02	1.000	27.0
1306.5	1513.5	207.0	5.684e-04	5.131e-09	3.496e+00	1.689e-02	1.000	26.3
Total			1.414e+04	5.035e-05	1.296e+08			

$D_{v0.1}$ = 90 $D_{v0.5}$ = 245 $D_{v0.9}$ = 414 (microns) Mass balance closure 89%

Atomizer: D8-46; Orientation: 90 degrees
Flowrate: 7.0 l/min; Windspeed: 45 m/s

Fluid: 9% isopropanol in water
Surface tension 40 dynes/cm

Lower Diameter (microns)	Upper Diameter (microns)	nDD	No. of Drops Noi (No.)	Volumetric Rate Fi (m^3/s)	Droplet Number Rate Ni (No./s)	Ni/DD	CVF	Calc. Droplet Velocity (m/s)
3.0	11.6	8.6	1.679e+05	2.199e-07	1.079e+09	1.255e+08	0.002	55.0
11.6	13.4	1.8	1.491e+04	9.806e-08	9.589e+07	5.327e+07	0.003	55.0
13.4	15.6	2.2	1.246e+04	1.279e-07	8.014e+07	3.643e+07	0.004	55.0
15.6	18.1	2.5	8.812e+03	1.419e-07	5.666e+07	2.267e+07	0.005	55.0
18.1	20.9	2.8	8.362e+03	2.084e-07	5.368e+07	1.917e+07	0.007	54.9
20.9	24.3	3.4	8.976e+03	3.476e-07	5.751e+07	1.691e+07	0.010	54.8
24.3	28.1	3.8	7.387e+03	4.449e-07	4.724e+07	1.243e+07	0.014	54.7
28.1	32.6	4.5	5.869e+03	5.464e-07	3.733e+07	8.295e+06	0.019	54.4
32.6	37.8	5.2	4.682e+03	6.763e-07	2.961e+07	5.695e+06	0.025	54.1
37.8	43.8	6.0	4.310e+03	9.624e-07	2.706e+07	4.510e+06	0.034	53.7
43.8	50.3	6.5	4.320e+03	1.471e-06	2.697e+07	4.156e+06	0.047	53.4
50.3	61.0	10.7	3.914e+03	2.187e-06	2.425e+07	2.269e+06	0.067	53.0
61.0	73.3	12.3	2.825e+03	2.745e-06	1.734e+07	1.413e+06	0.091	52.5
73.3	87.6	14.4	2.070e+03	3.437e-06	1.261e+07	8.766e+05	0.122	52.1
87.6	104.3	16.6	1.384e+03	3.862e-06	8.349e+06	5.021e+05	0.157	51.6
104.3	123.1	18.9	9.966e+02	4.592e-06	5.966e+06	3.161e+05	0.198	51.1
123.1	145.6	22.4	7.824e+02	5.889e-06	4.638e+06	2.067e+05	0.251	50.5
145.6	172.0	26.4	6.650e+02	8.162e-06	3.895e+06	1.476e+05	0.324	49.9
172.0	201.0	29.0	5.237e+02	1.031e-05	3.037e+06	1.046e+05	0.417	49.4
201.0	236.6	35.6	3.945e+02	1.242e-05	2.265e+06	6.356e+04	0.529	48.8
236.6	276.2	39.6	2.499e+02	1.251e-05	1.417e+06	3.579e+04	0.641	48.2
276.2	323.7	47.5	1.726e+02	1.366e-05	9.667e+05	2.035e+04	0.764	47.5
323.7	377.8	54.1	9.730e+01	1.216e-05	5.381e+05	9.945e+03	0.873	46.8
377.8	439.9	62.0	3.727e+01	7.282e-06	2.035e+05	3.281e+03	0.938	46.1
439.9	512.4	72.6	1.256e+01	3.825e-06	6.768e+04	9.324e+02	0.973	45.3
512.4	596.9	84.5	3.893e+00	1.850e-06	2.071e+04	2.452e+02	0.989	44.5
596.9	694.6	97.7	1.094e+00	8.079e-07	5.731e+03	5.868e+01	0.996	43.7
694.6	808.0	113.5	1.123e-01	1.285e-07	5.788e+02	5.100e+00	0.998	42.9
808.0	938.7	130.6	3.709e-02	6.565e-08	1.882e+02	1.441e+00	0.998	42.0
938.7	1091.8	153.1	2.814e-02	7.697e-08	1.405e+02	9.177e-01	0.999	41.1
1091.8	1267.3	175.5	1.482e-02	6.239e-08	7.261e+01	4.137e-01	0.999	40.3
1267.3	1471.8	204.5	1.026e-02	6.634e-08	4.932e+01	2.411e-01	1.000	39.3
Total			2.621e+05	1.114e-04	1.677e+09			

$D_{v0.1} = 77$ $D_{v0.5} = 227$ $D_{v0.9} = 404$ Mass balance closure 96%

Atomizer: D10-46; Orientation: 90 degress
Flowrate: 7.0 l/min; Windspeed: 45 m/s
Fluid: 9% isopropanol in water
Surafce tension 40 dynes/cm

Lower Diameter (microns)	Upper Diameter (microns)	DD	No. of Drops Noi (No.)	Volumetric Rate Fi (m^3/s)	Droplet Number Rate Ni (No./s)	Ni/DD	CVF	Calc. Droplet Velocity (m/s)
3.0	11.6	8.6	5.795e+04	1.518e-07	7.452e+08	8.666e+07	0.001	55.0
11.6	13.4	1.8	9.667e+03	1.271e-07	1.243e+08	6.907e+07	0.002	55.0
13.4	15.6	2.2	5.115e+03	1.050e-07	6.578e+07	2.990e+07	0.003	55.0
15.6	18.1	2.5	4.267e+03	1.375e-07	5.488e+07	2.195e+07	0.004	55.0
18.1	20.9	2.8	2.994e+03	1.492e-07	3.844e+07	1.373e+07	0.006	54.9
20.9	24.3	3.4	3.530e+03	2.734e-07	4.524e+07	1.330e+07	0.008	54.8
24.3	28.1	3.8	2.798e+03	3.370e-07	3.579e+07	9.419e+06	0.011	54.7
28.1	32.6	4.5	2.303e+03	4.288e-07	2.929e+07	6.509e+06	0.015	54.4
32.6	37.8	5.2	1.804e+03	5.212e-07	2.282e+07	4.389e+06	0.019	54.1
37.8	43.8	6.0	1.652e+03	7.378e-07	2.075e+07	3.458e+06	0.025	53.7
43.8	50.3	6.5	1.453e+03	9.890e-07	1.814e+07	2.795e+06	0.034	53.4
50.3	61.0	10.7	1.279e+03	1.430e-06	1.585e+07	1.483e+06	0.046	53.0
61.0	73.3	12.3	1.001e+03	1.945e-06	1.229e+07	1.001e+06	0.063	52.5
73.3	87.6	14.4	6.772e+02	2.249e-06	8.250e+06	5.735e+05	0.082	52.1
87.6	104.3	16.6	4.584e+02	2.559e-06	5.532e+06	3.327e+05	0.104	51.6
104.3	123.1	18.9	3.578e+02	3.296e-06	4.283e+06	2.270e+05	0.132	51.1
123.1	145.6	22.4	3.023e+02	4.551e-06	3.584e+06	1.597e+05	0.171	50.5
145.6	172.0	26.4	2.399e+02	5.889e-06	2.810e+06	1.065e+05	0.221	49.9
172.0	201.0	29.0	1.930e+02	7.599e-06	2.238e+06	7.709e+04	0.286	49.4
201.0	236.6	35.6	1.718e+02	1.082e-05	1.973e+06	5.536e+04	0.379	48.8
236.6	276.2	39.6	1.303e+02	1.304e-05	1.477e+06	3.731e+04	0.490	48.2
276.2	323.7	47.5	1.059e+02	1.676e-05	1.186e+06	2.497e+04	0.633	47.5
323.7	377.8	54.1	6.800e+01	1.700e-05	7.521e+05	1.390e+04	0.779	46.8
377.8	439.9	62.0	3.044e+01	1.189e-05	3.324e+05	5.359e+03	0.880	46.1
439.9	512.4	72.6	1.201e+01	7.321e-06	1.295e+05	1.784e+03	0.943	45.3
512.4	596.9	84.5	4.045e+00	3.845e-06	4.304e+04	5.095e+02	0.976	44.5
596.9	694.6	97.7	1.015e+00	1.499e-06	1.064e+04	1.089e+02	0.989	43.7
694.6	808.0	113.5	2.532e-01	5.797e-07	2.611e+03	2.300e+01	0.994	42.9
808.0	938.7	130.6	9.000e-02	3.186e-07	9.134e+02	6.991e+00	0.996	42.0
938.7	1091.8	153.1	3.440e-02	1.882e-07	3.435e+02	2.244e+00	0.998	41.1
1091.8	1267.3	175.5	1.448e-02	1.219e-07	1.419e+02	8.083e-01	0.999	40.3
1267.3	1471.8	204.5	8.713e-03	1.126e-07	8.374e+01	4.094e-01	1.000	39.3
Total			9.856e+04	1.170e-04	1.261e+09			

$D_{v0.1} = 101$ $D_{v0.5} = 280$ $D_{v0.9} = 463$ (microns) Mass balance closure 100%

Atomizer: D10-46; Orientation: 45 degrees
Flowrate: 7.0 l/min; Windspeed: 45 m/s

Fluid: 9% isopropanol in water
Surface tension 40 dynes/cm

Lower Diameter (microns)	Upper Diameter (microns)	DD	No. of Drops Noi (No.)	Volumetric Rate Fi (m^3/s)	Droplet Number Rate Ni (No./s)	Ni/DD	CVF	Calc. Droplet Velocity (m/s)
3.0	11.6	8.6	4.637e+04	1.215e-07	5.964e+08	6.934e+07	0.001	55.0
11.6	13.4	1.8	7.826e+03	1.029e-07	1.007e+08	5.592e+07	0.002	55.0
13.4	15.6	2.2	4.288e+03	8.803e-08	5.515e+07	2.507e+07	0.003	55.0
15.6	18.1	2.5	0.000e+00	0.000e+00	0.000e+00	0.000e+00	0.003	55.0
18.1	20.9	2.8	2.242e+03	1.117e-07	2.878e+07	1.028e+07	0.004	54.9
20.9	24.3	3.4	1.634e+03	1.265e-07	2.093e+07	6.157e+06	0.005	54.8
24.3	28.1	3.8	1.231e+03	1.482e-07	1.574e+07	4.143e+06	0.006	54.7
28.1	32.6	4.5	1.396e+03	2.600e-07	1.776e+07	3.946e+06	0.008	54.4
32.6	37.8	5.2	1.121e+03	3.238e-07	1.418e+07	2.726e+06	0.011	54.1
37.8	43.8	6.0	6.888e+02	3.076e-07	8.649e+06	1.441e+06	0.013	53.7
43.8	50.3	6.5	7.265e+02	4.946e-07	9.072e+06	1.398e+06	0.018	53.4
50.3	61.0	10.7	6.006e+02	6.711e-07	7.444e+06	6.964e+05	0.023	53.0
61.0	73.3	12.3	4.606e+02	8.951e-07	5.654e+06	4.607e+05	0.031	52.5
73.3	87.6	14.4	3.586e+02	1.191e-06	4.369e+06	3.037e+05	0.041	52.1
87.6	104.3	16.6	2.848e+02	1.590e-06	3.437e+06	2.067e+05	0.054	51.6
104.3	123.1	18.9	2.062e+02	1.900e-06	2.469e+06	1.308e+05	0.071	51.1
123.1	145.6	22.4	1.612e+02	2.426e-06	1.911e+06	8.517e+04	0.091	50.5
145.6	172.0	26.4	1.236e+02	3.034e-06	1.448e+06	5.485e+04	0.117	49.9
172.0	201.0	29.0	1.137e+02	4.476e-06	1.318e+06	4.541e+04	0.155	49.4
201.0	236.6	35.6	1.157e+02	7.287e-06	1.328e+06	3.728e+04	0.216	48.8
236.6	276.2	39.6	8.558e+01	8.568e-06	9.705e+05	2.451e+04	0.289	48.2
276.2	323.7	47.5	6.102e+01	9.660e-06	6.835e+05	1.439e+04	0.371	47.5
323.7	377.8	54.1	4.453e+01	1.113e-05	4.925e+05	9.103e+03	0.465	46.8
377.8	439.9	62.0	3.358e+01	1.312e-05	3.667e+05	5.912e+03	0.576	46.1
439.9	512.4	72.6	2.362e+01	1.439e-05	2.546e+05	3.508e+03	0.698	45.3
512.4	596.9	84.5	1.381e+01	1.313e-05	1.469e+05	1.739e+03	0.809	44.5
596.9	694.6	97.7	6.366e+00	9.402e-06	6.669e+04	6.829e+02	0.889	43.7
694.6	808.0	113.5	2.565e+00	5.874e-06	2.645e+04	2.331e+02	0.938	42.9
808.0	938.7	130.6	1.037e+00	3.672e-06	1.053e+04	8.059e+01	0.969	42.0
938.7	1091.8	153.1	3.865e-01	2.115e-06	3.860e+03	2.521e+01	0.987	41.1
1091.8	1267.3	175.5	1.233e-01	1.038e-06	1.208e+03	6.881e+00	0.996	40.3
1267.3	1471.8	204.5	3.505e-02	4.530e-07	3.368e+02	1.647e+00	1.000	39.3
Total			7.022e+04	1.181e-04	8.997e+08			

$D_{v0.1}$ = 155 $D_{v0.5}$ = 397 $D_{v0.9}$ = 720 (microns) Mass balance closure 101%

Atomizer: D10-46; Orientation: 90 degrees
Flowrate: 7.0 l/min; Windspeed: 36 m/s

Fluid: 9% isopropanol in water
Surface tension 40 dynes/cm

Lower Diameter (microns)	Upper Diameter (microns)	DD	No. of Drops Noi (No.)	Volumetric Rate Fi (m^3/s)	Droplet Number Rate Ni (No./s)	Ni/DD	CVF	Calc. Droplet Velocity (m/s)
3.0	11.6	8.6	4.411e+04	7.564e-08	3.714e+08	4.318e+07	0.001	36.0
11.6	13.4	1.8	1.067e+04	9.188e-08	8.984e+07	4.991e+07	0.001	36.0
13.4	15.6	2.2	7.173e+03	9.638e-08	6.038e+07	2.745e+07	0.002	36.0
15.6	18.1	2.5	3.631e+03	7.657e-08	3.057e+07	1.223e+07	0.003	36.0
18.1	20.9	2.8	2.957e+03	9.665e-08	2.489e+07	8.890e+06	0.004	36.0
20.9	24.3	3.4	3.161e+03	1.608e-07	2.661e+07	7.825e+06	0.005	36.0
24.3	28.1	3.8	2.488e+03	1.972e-07	2.094e+07	5.511e+06	0.006	36.0
28.1	32.6	4.5	2.056e+03	2.534e-07	1.731e+07	3.847e+06	0.008	36.0
32.6	37.8	5.2	1.745e+03	3.354e-07	1.469e+07	2.825e+06	0.011	36.0
37.8	43.8	6.0	1.656e+03	4.944e-07	1.390e+07	2.317e+06	0.015	35.9
43.8	50.3	6.5	1.468e+03	6.680e-07	1.225e+07	1.888e+06	0.021	35.7
50.3	61.0	10.7	1.193e+03	8.927e-07	9.901e+06	9.262e+05	0.028	35.5
61.0	73.3	12.3	8.161e+02	1.066e-06	6.737e+06	5.489e+05	0.036	35.3
73.3	87.6	14.4	6.086e+02	1.354e-06	4.967e+06	3.453e+05	0.047	34.9
87.6	104.3	16.6	4.624e+02	1.730e-06	3.741e+06	2.250e+05	0.061	34.6
104.3	123.1	18.9	3.790e+02	2.333e-06	3.031e+06	1.606e+05	0.080	34.2
123.1	145.6	22.4	2.819e+02	2.838e-06	2.235e+06	9.961e+04	0.103	33.9
145.6	172.0	26.4	1.841e+02	3.022e-06	1.442e+06	5.464e+04	0.128	33.5
172.0	201.0	29.0	1.507e+02	3.960e-06	1.166e+06	4.017e+04	0.160	33.1
201.0	236.6	35.6	1.528e+02	6.391e-06	1.165e+06	3.270e+04	0.211	32.6
236.6	276.2	39.6	1.239e+02	8.208e-06	9.298e+05	2.348e+04	0.278	32.1
276.2	323.7	47.5	9.514e+01	9.936e-06	7.030e+05	1.480e+04	0.358	31.6
323.7	377.8	54.1	7.329e+01	1.201e-05	5.313e+05	9.819e+03	0.455	31.0
377.8	439.9	62.0	6.055e+01	1.545e-05	4.318e+05	6.962e+03	0.580	30.5
439.9	512.4	72.6	4.636e+01	1.832e-05	3.241e+05	4.466e+03	0.728	29.9
512.4	596.9	84.5	2.575e+01	1.571e-05	1.758e+05	2.081e+03	0.855	29.2
596.9	694.6	97.7	8.399e+00	7.919e-06	5.617e+04	5.752e+02	0.919	28.6
694.6	808.0	113.5	2.214e+00	3.207e-06	1.444e+04	1.273e+02	0.945	27.9
808.0	938.7	130.6	1.086e+00	2.410e-06	6.908e+03	5.287e+01	0.965	27.2
938.7	1091.8	153.1	5.434e-01	1.845e-06	3.367e+03	2.199e+01	0.980	26.5
1091.8	1267.3	175.5	2.730e-01	1.415e-06	1.647e+03	9.385e+00	0.991	25.8
1267.3	1471.8	204.5	1.399e-01	1.100e-06	8.177e+02	3.998e+00	1.000	25.0
Total			8.579e+04	1.237e-04	7.203e+08			

$D_{v0.1}$ = 143 $D_{v0.5}$ = 400 $D_{v0.9}$ = 666 (microns) Mass balance closure 106%

Atomizer: D10-46; Orientation: 135 degrees
Flowrate: 7.0 l/min; Windspeed: 45 m/s

Fluid: 9% isopropanol in water
Surface tension 40 dynes/cm

Lower Diameter (microns)	Upper Diameter (microns)	DD	No. of Drops Noi (No.)	Volumetric Rate Fi (m^3/s)	Droplet Number Rate Ni (No./s)	Ni/DD	CVF	Calc. Droplet Velocity (m/s)
3.0	11.6	8.6	1.487e+05	3.897e-07	1.913e+09	2.224e+08	0.004	55.0
11.6	13.4	1.8	1.785e+04	2.348e-07	2.296e+08	1.276e+08	0.007	55.0
13.4	15.6	2.2	1.248e+04	2.562e-07	1.605e+08	7.295e+07	0.009	55.0
15.6	18.1	2.5	7.907e+03	2.547e-07	1.017e+08	4.067e+07	0.012	55.0
18.1	20.9	2.8	7.320e+03	3.648e-07	9.397e+07	3.356e+07	0.016	54.9
20.9	24.3	3.4	7.501e+03	5.810e-07	9.612e+07	2.827e+07	0.022	54.8
24.3	28.1	3.8	6.731e+03	8.107e-07	8.609e+07	2.266e+07	0.030	54.7
28.1	32.6	4.5	5.546e+03	1.033e-06	7.054e+07	1.568e+07	0.041	54.4
32.6	37.8	5.2	4.456e+03	1.287e-06	5.637e+07	1.084e+07	0.055	54.1
37.8	43.8	6.0	3.474e+03	1.551e-06	4.362e+07	7.270e+06	0.071	53.7
43.8	50.3	6.5	2.963e+03	2.017e-06	3.700e+07	5.701e+06	0.092	53.4
50.3	61.0	10.7	2.362e+03	2.639e-06	2.927e+07	2.738e+06	0.119	53.0
61.0	73.3	12.3	1.697e+03	3.298e-06	2.084e+07	1.698e+06	0.154	52.5
73.3	87.6	14.4	1.177e+03	3.908e-06	1.434e+07	9.966e+05	0.195	52.1
87.6	104.3	16.6	7.791e+02	4.348e-06	9.401e+06	5.654e+05	0.240	51.6
104.3	123.1	18.9	5.445e+02	5.018e-06	6.520e+06	3.455e+05	0.293	51.2
123.1	145.6	22.4	4.087e+02	6.153e-06	4.846e+06	2.160e+05	0.357	50.7
145.6	172.0	26.4	3.071e+02	7.538e-06	3.597e+06	1.363e+05	0.436	50.1
172.0	201.0	29.0	2.130e+02	8.387e-06	2.470e+06	8.508e+04	0.524	49.6
201.0	236.6	35.6	1.516e+02	9.545e-06	1.740e+06	4.884e+04	0.624	49.1
236.6	276.2	39.6	9.658e+01	9.670e-06	1.095e+06	2.767e+04	0.725	48.5
276.2	323.7	47.5	6.447e+01	1.021e-05	7.221e+05	1.520e+04	0.831	47.9
323.7	377.8	54.1	3.112e+01	7.779e-06	3.442e+05	6.361e+03	0.913	47.3
377.8	439.9	62.0	7.746e+00	3.027e-06	8.459e+04	1.364e+03	0.944	46.7
439.9	512.4	72.6	2.643e+00	1.611e-06	2.850e+04	3.926e+02	0.961	46.1
512.4	596.9	84.5	1.231e+00	1.170e-06	1.310e+04	1.551e+02	0.974	45.5
596.9	694.6	97.7	4.740e-01	7.001e-07	4.966e+03	5.085e+01	0.981	44.8
694.6	808.0	113.5	1.825e-01	4.179e-07	1.882e+03	1.658e+01	0.985	44.1
808.0	938.7	130.6	9.815e-02	3.474e-07	9.960e+02	7.624e+00	0.989	43.4
938.7	1091.8	153.1	6.361e-02	3.480e-07	6.351e+02	4.149e+00	0.993	42.7
1091.8	1267.3	175.5	4.264e-02	3.590e-07	4.178e+02	2.380e+00	0.996	41.9
1267.3	1471.8	204.5	2.737e-02	3.538e-07	2.631e+02	1.286e+00	1.000	41.1
Total			2.328e+05	9.560e-05	2.984e+09			

$D_{v0.1}$ = 53 $D_{v0.5}$ = 193 $D_{v0.9}$ = 369 (microns) Mass balance closure 82%

Atomizer: D8-46; Orientation: 90 degrees
Flowrate: 3.7 l/min; Windspeed: 45 m/s

Fluid: 9% isopropanol in water
Surface tension 40 dynes/cm

Lower Diameter (microns)	Upper Diameter (microns)	DD	No. of Drops Noi (No.)	Volumetric Rate Fi (m^3/s)	Droplet Number Rate Ni (No./s)	Ni/DD	CVF	Calc. Droplet Velocity (m/s)
3.0	11.6	8.6	5.437e+04	1.424e-07	6.993e+08	8.131e+07	0.002	55.0
11.6	13.4	1.8	7.252e+03	9.538e-08	9.326e+07	5.181e+07	0.003	55.0
13.4	15.6	2.2	4.929e+03	1.012e-07	6.339e+07	2.881e+07	0.005	55.0
15.6	18.1	2.5	2.336e+03	7.524e-08	3.004e+07	1.201e+07	0.006	55.0
18.1	20.9	2.8	2.196e+03	1.094e-07	2.819e+07	1.007e+07	0.008	54.9
20.9	24.3	3.4	2.053e+03	1.590e-07	2.631e+07	7.738e+06	0.010	54.8
24.3	28.1	3.8	2.023e+03	2.437e-07	2.588e+07	6.810e+06	0.014	54.7
28.1	32.6	4.5	1.552e+03	2.890e-07	1.974e+07	4.387e+06	0.018	54.4
32.6	37.8	5.2	1.179e+03	3.406e-07	1.491e+07	2.868e+06	0.023	54.1
37.8	43.8	6.0	1.084e+03	4.841e-07	1.361e+07	2.269e+06	0.030	53.7
43.8	50.3	6.5	1.034e+03	7.042e-07	1.292e+07	1.990e+06	0.040	53.4
50.3	61.0	10.7	7.966e+02	8.902e-07	9.873e+06	9.236e+05	0.053	53.0
61.0	73.3	12.3	5.848e+02	1.136e-06	7.179e+06	5.850e+05	0.070	52.5
73.3	87.6	14.4	4.028e+02	1.338e-06	4.907e+06	3.411e+05	0.089	52.1
87.6	104.3	16.6	2.667e+02	1.488e-06	3.217e+06	1.935e+05	0.111	51.6
104.3	123.1	18.9	2.156e+02	1.982e-06	2.576e+06	1.365e+05	0.140	51.1
123.1	145.6	22.4	1.855e+02	2.781e-06	2.190e+06	9.764e+04	0.181	50.5
145.6	172.0	26.4	1.543e+02	3.772e-06	1.800e+06	6.821e+04	0.236	49.9
172.0	201.0	29.0	1.207e+02	4.733e-06	1.394e+06	4.801e+04	0.305	49.4
201.0	236.6	35.6	1.061e+02	6.641e-06	1.211e+06	3.397e+04	0.402	48.8
236.6	276.2	39.6	8.065e+01	8.025e-06	9.090e+05	2.296e+04	0.519	48.2
276.2	323.7	47.5	6.296e+01	9.883e-06	6.993e+05	1.472e+04	0.663	47.5
323.7	377.8	54.1	3.822e+01	9.452e-06	4.182e+05	7.729e+03	0.802	46.8
377.8	439.9	62.0	1.802e+01	6.952e-06	1.943e+05	3.132e+03	0.903	46.1
439.9	512.4	72.6	6.691e+00	4.006e-06	7.088e+04	9.765e+02	0.962	45.3
512.4	596.9	84.5	2.011e+00	1.869e-06	2.092e+04	2.477e+02	0.989	44.5
596.9	694.6	97.7	3.977e-01	5.729e-07	4.064e+03	4.161e+01	0.997	43.7
694.6	808.0	113.5	6.794e-02	1.513e-07	6.815e+02	6.005e+00	1.000	42.9
808.0	938.7	130.6	5.083e-03	1.741e-08	4.992e+01	3.821e-01	1.000	42.0
938.7	1091.8	153.1	2.018e-03	1.063e-08	1.940e+01	1.267e-01	1.000	41.1
1091.8	1267.3	175.5	2.552e-04	2.067e-09	2.405e+00	1.370e-02	1.000	40.3
1267.3	1471.8	204.5	8.152e-05	1.008e-09	7.491e-01	3.662e-03	1.000	39.3
Total			8.305e+04	6.845e-05	1.064e+09			

$D_{v0.1}$ = 96 $D_{v0.5}$ = 270 $D_{v0.9}$ = 438 Mass balance closure 111%

APPENDIX III

DETAILS OF AGDISP, FSCBG, AND PKBW MODELS

ITEM	AGDISP	FSCBG	PKBW
INPUT: METEOROLOGY	Neutral atm. stab'y, uses velocity temp., rel. hum., turbulence	As for AGDISP	All stability classes, uses friction velocity, temp., rel. hum., Monin-Obukhov length
AIRCRAFT WAKE	Decaying Lamb vortex, uses a/c weight, wingspan, speed, altitude	Uses AGDISP output for near-wake calculations	Nasa WAKE VORTEX code for calc'n of decaying vortex (using wind velocity and temp. gradient, a/c weight, speed, wingspan, altitude)
DROPLET SPECTRUM	16 size classes, convertible to 100 by curve-fitting, nozzles distributed on booms	As AGDISP	7 size classes, additional resolution by repeating simulation, nozzles distributed on booms
EVAPORATION	Trayford and Welch (1977) method [J.Agr.Engr.Res. 22:183-196]	As AGDISP	Picot, Chitrangad, Henderson (1981) method [Trans. ASAE 24:552-554]
CANOPY DEPOSIT	Tree envelope, probability of penetration, deposition on canopy and ground	As AGDISP	Vertical distribution of foliage density, probabilities of interaction and deposition each trajectory step, collection efficiency for foliage type

CALCULATION METHOD:	Mean trajectories calculated with dispersion about the mean	AGDISP near wake results are passed to FSCBG as initiator, then Gaussian line-source theory is used; any orientation of line to wind direction, multi-swath plot spraying included	Trajectory calculation for 1000 droplets per size class, subsequently normalized for atomizer spectrum. Plot spraying done by spread-sheet
OUTPUT:	Graphical outputs, tables of data	Complete plotting for inputs and results (e.g. deposit over entire spray plot), tables of data	Limited direct plotting, provides tables of drift cloud and canopy & ground deposits, also drops/needle or drops/leaf with size dist'n, file outputs to spreadsheets
COMPUTER:	IBM PC and compatibles, USDA mainframe system	As AGDISP	IBM PC and compatibles, 486 or higher recommended
VALIDATION:	USDA Forest Service report FPM 90-1, Forest Pest Management, Davis, CA. "AGDISP Evaluation and validation training" by M. Teske	Envir. Toxicology & Chem. **12**, (1993) pp 1977-1989, "Predicting and measuring drift of Bt sprays", by Barry, Skyler, and Teske	Agr. And For. Met'y **76**, (1994), pp19-40, "A numerical model for forestry aerial spraying", by Wallace, Picot, and Chapman
AVAILABILITY:	M.E. Teske, Continuum Dynamics, Inc., P.O. Box 3073, Princeton, N.J. 08543.	As AGDISP	Effective Particle Technologies Ltd., 29 Simcoe Ct., Fredericton, N.B. Can. E3B2W9

INDEX

Environmental Science and Technology Library

1. A. Caetano, M.N. De Pinho, E. Drioli and H. Muntau (eds.), *Membrane Technology: Applications to Industrial Wastewater Treatment.* 1995 ISBN 0-7923-3209-1
2. Z. Zlatev: *Computer Treatment of Large Air Pollution Models.* 1995 ISBN 0-7923-3328-4
3. J. Lemons and D.A. Brown (eds.): *Sustainable Development: Science, Ethics, and Public Policy.* 1995 ISBN 0-7923-3500-7
4. A.V. Gheorghe and M. Nicolet-Monnier: *Integrated Regional Risk Assessment.* Volume I: Continuous and Non-Point Source Emissions: Air, Water, Soil. 1995 ISBN 0-7923-3717-4
   Volume II: Consequence Assessment of Accidental Releases. 1995 ISBN 0-7923-3718-2
   Set: ISBN 0-7923-3719-0
5. L. Westra and J. Lemons (eds.): *Perspectives on Ecological Integrity.* 1995 ISBN 0-7923-3734-4
6. J. Sathaye and S. Meyers: *Greenhouse Gas Mitigation Assessment: A Guidebook.* 1995 ISBN 0-7923-3781-6
7. R. Benioff, S. Guill and J. Lee (eds.): *Vulnerability and Adaptation Assessments.* An International Handbook. 1996 ISBN 0-7923-4140-6
8. J.B. Smith, S. Huq, S. Lenhart, L.J. Mata, I. Nemošová and S. Toure (eds.): *Vulnerability and Adaptation to Climate Change.* Interim Results from the U.S. Country Studies Program. 1996 ISBN 0-7923-4141-4
9. B.V. Braatz, B.P. Jallow, S. Molnár, D. Murdiyarso, M. Perdomo and J.F. Fitzgerald (eds.): *Greenhouse Gas Emission Inventories.* Interim Results from the U.S. Country Studies Program. 1996 ISBN 0-7923-4142-2
10. M. Palo and G. Mery (eds.): *Sustainable Forestry Challenges for Developing Countries.* 1996 ISBN 0-7923-3738-7
11. S. Guerzoni and R. Chester (eds.): *The Impact of Desert Dust Across the Mediterranean.* 1996 ISBN 0-7923-4294-1
12. J.J.C. Picot and D.D. Kristmanson: *Forestry Pesticide Aerial Spraying.* Spray Droplet Generation, Dispersion, and Deposition. 1997 ISBN 0-7923-4371-9

KLUWER ACADEMIC PUBLISHERS – DORDRECHT / BOSTON / LONDON

Kolibri 360 GmbH
Ferdinand-Jühlke-Straße 7
99095 Erfurt, Deutschland
produktsicherheit@kolibri360.de